Reviews and critical articles covering the entire field of normal anatomy (cytology, histology, cyto- and histochemistry, electron microscopy, macroscopy, experimental morphology and embryology and comparative anatomy) are published in Advances in Anatomy, Embryology and Cell Biology. Papers dealing with anthropology and clinical morphology that aim to encourage co-operation between anatomy and related disciplines will also be accepted. Papers are normally commissioned. Original papers and communications may be submitted and will be considered for publication provided they meet the requirements of a review article and thus fit into the scope of "Advances". English language is preferred, but in exceptional cases French or German papers will be accepted.

Manuscripts should be addressed to

Prof. Dr. A. **BRODAL,** Universitetet i Oslo, Anatomisk Institutt, Karl Johans Gate 47 (Domus Media), Oslo 1/Norway

Prof. W. **HILD,** Department of Anatomy, Medical Branch, The University of Texas, Galveston, Texas 77550/USA

Prof. Dr. J. van **LIMBORGH,** Universiteit van Amsterdam, Anatomisch-Embryologisch Laboratorium, Mauritskade 61, Amsterdam-O/Holland

Prof. Dr. R. **ORTMANN,** Anatomisches Institut der Universität, Lindenburg, D-5000 Köln-Lindenthal

Prof. Dr. T. H. **SCHIEBLER,** Anatomisches Institut der Universität, Koellikerstraße 6, D-8700 Würzburg

Prof. Dr. G. **TÖNDURY,** Direktion der Anatomie, Gloriastraße 19, CH-8006 Zürich/Schweiz

Prof. Dr. E. **WOLFF,** Collège de France, Laboratoire d'Embryologie Expérimentale, 11, Place Marcelin Berthelot, F-75005 Paris/France

Advances in Anatomy
Embryology and Cell Biology

Vol. 59

Editors
A. Brodal, Oslo   W. Hild, Galveston
J. van Limborgh, Amsterdam
R. Ortmann, Köln   T. H. Schiebler, Würzburg
G. Töndury, Zürich   E. Wolff, Paris

Thomas Bär

The Vascular System of the Cerebral Cortex

With 33 Figures

Springer-Verlag
Berlin Heidelberg New York 1980

Priv.-Doz. Dr. Thomas Bär
Max-Planck-Institut für Systemphysiologie
Rheinlanddamm 201, 4600 Dortmund 1

Habilitationsschrift, translated from the German and published
with the permission of the Medizinische Fakultät an der Rhein.-
Westf. Techn. Hochschule Aachen.
German Title: Morphometrische Untersuchungen am Gefäß-
system der Hirnrinde. Ontogenese – Altersveränderungen – An-
passung an Hypoxie.

ISBN-13: 978-3-540-09652-8   e-ISBN-13: 978-3-642-67432-7
DOI: 10.1007/978-3-642-67432-7

Library of Congress Cataloging in Publication Data. Bär, Thomas,
1943– The vascular system of the cerebral cortex. (Advances in anatomy, embryology, and cell biology; v. 59) Bibliography: p. Includes
index. 1. Cerebral cortex – Blood-vessels. I. Title. II. Series. QL801.
E67 vol. 59 [QM455] 574.4'08s [612'.825] ISBN 0-387-09652-3
79-23217

Composition: SatzStudio Pfeifer, Germering

2121/3321-543210

Contents

Acknowledgements

An essential part of the experiments was carried out in the Department of Neurobiology of the Max-Planck-Institut für Biophysikalische Chemie in Göttingen from 1972 to 1975. I am very indebted to Prof. J.R. Wolff for his continuing interest and for many stimulating discussions. I would like to express my appreciation to Prof. W. Lange for his support which led to the accomplishment of the present work. I am grateful to Miss L. Strauch for her expert assistance, to Mr. W. Graulich for the drawing of graphs, and to Mrs. J. Jacobs for typewriting the manuscript. The financial support by the Deutsche Forschungsgemeinschaft (Ba 609/1) is gratefully acknowledged.

Abbreviations

L_E (μm)	mean length of endothelial cells of capillaries
L_K (μm)	mean length of endothelial nuclei of capillaries
L_V (mm/mm^3)	specific length of capillaries
L_{rad}	length of vertically oriented intracortical vascular trunks (classicication according to the depth of penetration)
mK	capillary cross-section with part of endothelial nucleus in section plane
N_{Arad} (N/mm^2)	mean packing-density of vertically oriented intracortical vascular trunks (number per mm^2 of cortical surface)
N_E (N/mm^3)	the mean numerical density of endothelial cells of capillaries (number per mm^3 of cortical tissue)
N_{ECJ}	number of interendothelial contact zones per capillary cross-section
N_{VY} (N/mm^3)	mean volume-density of vascular branchings
oK	capillary cross-section without part of endothelial nucleus in section plane
S_V (mm^2/mm^3)	specific internal surface area of capillaries
V_V (%)	volume fraction of capillaries (in percent of cortical volume)

1 Introduction

A vascular system consists of a supplying arterial and a draining venous part which are connected by a terminal vascular network. The arterial segment can be characterized according to the structural features of the vessel wall. However, it is sometimes difficult to distinguish the capillary from the postcapillary vessels on the basis of structural features alone. On the other hand, physiologic qualities such as permeability can hardly be associated with an equivalent histologic pattern of the vessel wall (Illig 1961; Rhodin 1967, 1968; Hauck 1971; Westergaard 1974). A definition of a vascular segment based on biologic significance should combine morphological and functional qualities of the vessel walls.

During the ontogeny of the mammalian organism a variety of vascular patterns (e.g., distribution of arteries and veins, arrangement of the capillaries) has been formed typical of each organ (Wolff et al. 1975; Baez 1977). The capillaries connect the feeding arterioles and the collecting venules in two different ways according to the branching pattern of the terminal vessels (Hauck 1975, Wolff et al., 1975). The arterioles and venules are directly connected by capillary segments. Consequently a terminal vessel called arteriovenous *(a-v) capillary* results, or a closely meshed *capillary network* is developed which connects arterioles and venules by a variable number of small capillary branches arranged parallel to the preexisting a-v capillary. The organotypic microvascular pattern is realized by a different representation of these two forms of arrangement of capillaries. The intracerebral microvascular system is characterized by a marked development of small netcapillaries so that the original terminal vessels connecting arteries and veins become difficult to distinguish (Wolff et al. 1975). The definitive angioarchitectonic is associated with the organ-specific morphogenesis and differentiation.

Historical survey. Thomas Willis' book *Cerebri Anatome,* which appeared in 1664 (cited by Bull 1975), can be regarded as a milestone in the history of modern neurology. He described the anastomosing ring of the basal intracranial arteries in mammals and denied the existence of a rete mirabile connecting the carotids with the intracranial arteries in man. In 1876 Vicq d' Azyr published excellent drawings demonstrating the course of the parietal branches of the intracranial arteries and the occurrence of venous anastomoses at the cerebral convexity. Duret (1874), an assistant of the famous Charcot, systematically studied the leptomeningeal and the intracortical vascular system. According to the different length of the vascular trunks that supply the cerebral cortex and those that supply the subjacent white matter, Duret (1874) was able to distinguish between rami corticales and rami medullares.

In 1874 Heubner examined the terminal branches of the leptomeningeal arteries and succeeded in demonstrating interarterial anastomoses. This observation disagreed with the results of Duret (1874) who denied the functional relevance of an arterial collateral circulation on the convex surfaces of the hemispheres. Today there is no doubt that the parietal branches of the leptomeningeal arteries are connected by arterial anastomoses (van den Bergh and van der Eecken 1968). The intracerebral branches of the leptomeningeal arteries have been regarded as end-arteries since the days of Cohnheim (1872) and Duret. During the last 50 years, however, many authors found anastomoses of various calibers connecting intracortical arteries and veins (Pfeifer 1928, 1931; Hasegawa et al. 1967, van den Bergh and van der Eecken 1968). There is no agreement about the frequency of occurrence, the anatomic localization, and the functional relevance.

Since Rosenthal's first description of the basal cerebral vein (1824, cited by Bull 1975) only few studies have been directed to the intracortical venous drainage (Kaplan and Ford 1966, Vuia 1966). Irrespective of the reports available, the venous side of the cerebrovascular system has been neglected in spite of its pathophysiologic importance.

The relationship between brain softening and pathologic alterations of brain arteries was emphasized by Rostan in 1823. In 1852 (cited by Bull 1975) a publication of W.S. 'Kirkes dealt with the relation between thrombembolic processes of intracranial vessels and the ensuing neurologic deficits. Thus, the importance of the regional cerebral angioarchitectonic (including local differences in number and arrangement of vessels) for the development and extension of a brain lesion was clearly recognized more than 100 years ago. Nevertheless, a correlation of the local vascular pattern to the size and distribution of a possible brain lesion remains difficult, especially if interpretation is attempted in terms of anatomy alone (e.g., the discrepancy between the size of an ischemic necrosis and that of the corresponding vascular territory after embolic occlusion of an intracortical artery).

Two hundred years after capillaries had been made visible for the first time (in the frog's lung) by Malpighi in 1661, Duret (1874) described the distribution of capillaries in different laminae of the cerebral cortex. However, the biologic significance of the differences in capillary density between various brain regions as pointed out by Pfeifer (1928) remains to be elucidated. In the early 1920s it became necessary to describe the capillary net by *quantitative methods* in order to achieve comparable results. Craigie started his comprehensive studies concerning morphometry of rat brain capillaries in 1920. Pfeifer, however, took a different way by using an improved vascular injection technique. He was able to show the three-dimensional arrangement of the intracortical vessels. The apparent differences in many cortical regions and laminae caused him to divide the cerebral cortex into several angioarchitectonic areas (Pfeifer 1940).

The *functional properties* of the complex three-dimensional arrangement of the microvessels can hardly be described by morphometric parameters like capillary length as used by Craigie (1920). The pattern of perfusion and the hemodynamic conditions are influenced by the number, the spatial distribution, and the angles of the vascular branchings (Lübbers 1968, Grunewald 1969, Baez 1977).

The regional specific topographical relationship between vessels and neuronal elements additionally plays an important role for a sufficient blood supply (Scharrer 1944, 1962; Balaschova 1956). The number of capillaries seems to depend on the local metabolic activity (Friede 1966, Diemer 1968). Presuming that the transport of substrates of the energy metabolism represents a predominating function of the vascular system, the density of capillaries may be correlated to the activity of oxidative enzymes and to the oxygen consumption in distinct brain regions (Campbell 1939, Friede 1966). This agrees with the correlation between the density of mitochondria and capillaries demonstrated by Scharrer (1945). There are additional factors that influence the regional supply with capillaries. In parts of the brain that show neurosecretory activity (e.g., some hypothalamic nuclei, periventricular regions) the local microcirculation serves the transport and the distribution of neurohormones.

Thus, regional differences in capillary density can only be understood in relation to the structural and functional organization of the brain tissue. It cannot generally be stated that a correlation between capillary density and a distinct neuronal structure exists, because the metabolic activity of morphologically equivalent structures (e.g., plasma membranes or synapses) may vary according to the functional specialization. The total capillary surface area, which depends on the length and the mean diameter of the vessels, represents an important morphometric parameter by which exchange processes between blood and brain tissue may be limited.

The understanding of the vascular system is improved by comparative studies of *phylogenic and ontogenic development*. A simple relation between the level of evolution and vascularity should not be expected because of various interactions between morphogenesis and differentition of the tissue on the one hand and vascularization on the other. A penetration of vessels into the brain tissue is considered as a step toward a higher level of evolution when different species of cyclostomes are compared (Spalteholz 1923). However, during phylogeny an increasing functional differentiation of the CNS neither runs parallel to an extension of the vascular system nor to an increased relative oxygen consumption. In the mammalian brain the relative oxygen consumption (measured by the activity of cytochrome oxidase or by the regional glucose utilization) is influenced by parameters such as the whole body size of the species under examination (Craigie 1945; Horstmann 1960; Friede 1966; Allen et al. 1977; Sokoloff 1977).

During ontogenic development the maturation of the brain is accompanied by an extending intracerebral vascularization (Feeney and Watterson 1946; Caley and Maxwell 1970). The brains of *altricial species* (such as mouse, rat, and cat, and including man) show a poor vascularization at birth; on the other hand, *precocious species* (e.g., guinea pig, chicken), which are relatively mature at term, are characterized by an advanced brain development showing a correspondingly dense capillary network (Horstmann 1959). The studies of ontogenic development suggest an interaction between the differentiation of neuronal metabolism (i.e., maturation of enzyme pattern) and the growth of capillaries. This interaction can be disturbed by the application of toxic drugs or by modification of the environmental conditions during development. Neonatal rats that were made cretinous by administration of methylthiouracil show a hypoplastic intracortical capillary bed accompanying the reduction of the metabolism of the nervous tissue (Eayrs 1954). Visual deprivation immediately after birth causes a retardation of neuronal growth and differentiation, as observed in mice (Gyllensten 1959) and monkey (Vital-Durand et al. 1978). Under these conditions the vascular density in the area striata has relatively increased after 20 days in darkness and decreases during the following development (Gyllensten 1959). The vascular growth appears to be secondarily influenced by the retardation of the differentiation of neurons. However, there is not enough evidence to decide whether an alteration of the vascular system is the cause or the consequence of the accompanying changes in the neuronal tissue (Eayrs 1954). A direct effect of drugs on the vascular growth cannot be excluded (Pentschew and Garro 1966, Ashton et al. 1972). Modification of the density of the intracerebral capillary net was observed after environmental enrichment and after increasing the motor performance (e.g., by tread mill running). In adult animals an increase in capillary density of the motor cortex was found after endurance training (Petrén 1938). After longlasting hypoxia, Diemer (1968) and Burian (1970) observed an increased density of cortical capillaries. However, it remains questionable whether an additional supply with capillaries is necessary or whether enzymatic adaptation is sufficient to meet the metabolic requirements of the increased neuronal activity.

One aim of the present study is to describe the successive ontogenic development of the intracortical vascular tree. The factors influencing vascular growth may change during different ontogenic periods. The topographical relations of the main vascular trunks seem to be determined by morphogenetic processes (Lierse 1963). The question arises which parts of the existing vascular system are able to adapt to the changes of functional activity (e.g., spontaneous activity, pattern of excitation and inhibition) that accompany the maturation of different regions of the brain. A quantitative evaluation of the intracortical vascular system must take into account all parts of the vascular tree. An attempt should be made to obtain a definition of the capillaries based upon objective criteria such as diameter and orientation distribution. New quantitative data concerning capillary surface area can be obtained, which may improve the understanding of developmental changes of the exchange processes between blood and brain tissue.

Based upon the morphometric data the process of vascularization can be mathematically simulated. Compared with the actual neuropathologic findings such mathematical models may help to explain potential disturbances of cortical architecture caused by developmental abnormalities of the vascular system.

In the experimental part of the present work the effects of a mild oxygen deprivation upon cortical capillaries are analyzed. Irrespective of the reports available it cannot be decided whether a growth of new capillaries (i.e., hyperplasia of endothelial cells) occurs under the conditions of oxygen deficiency.

Quantitative studies are applied to detect the age changes of the microcirculatory bed, which accompany the involution of the brain. The resulting morphological alterations of the capillary wall may be associated with the impairment of the elastic properties of the capillaries and of the transendothelial exchange processes. Abnormal regulation of the regional blood flow may be a further consequence of the limited plasticity of the terminal vessels. Thus, a quantitative approach to brain microvessels may contribute to knowledge of the cellular aspects of aging.

2 Material and Methods

This study is based on the brains of 137 Sprague Dawley and Wistar rats ranging in age from 11 days postcoitum (p.c.) to 30 months postpartum (p.p.). To obtain dated embryos the rats were kept at 24°C and 60% air humidity at a defined dark/light rhythm. As soon as sperm could be detected in the vaginal smears at the estrus stage this day was considered day 0 of pregnancy (E-0).

Table 1. Details of animals examined

Age (days after birth)	No. of animals injected with Indian ink	No. of animals fixed by perfusion and embedded in Epon (in brackets, ^{3}H-thymidine autoradiography)
1	2	
2/5	6	6 (5)
7	–	2 (2)
8	2	6 (5)
11/12	3	2 (2)
13/15	6	8 (5)
20	5	5 (2)
30	10	6 (6)
55	–	6
70	3	8 (3)
120	5	6
180	3	4
330	–	2
550	–	3
690	–	4
900	–	4

4

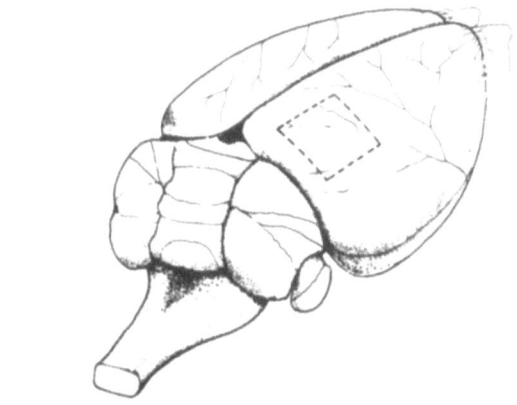

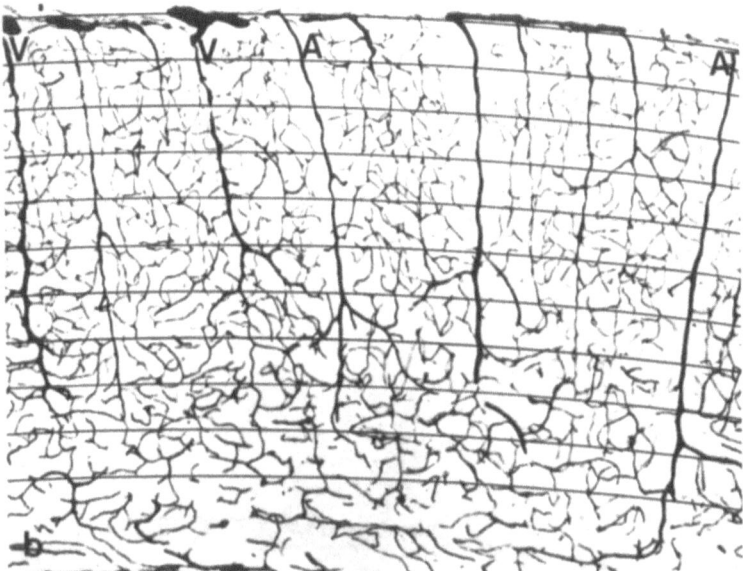

Fig. 1 a. View of a rat's brain. The area marked with *broken lines* shows the examined part of the cortex; b. Frontal (coronal) section (100 μm thick) through the cortex of a 14-day-old rat. Note the different lenghts of arteries (*A*) and veins. (*V*). x 50,4

The first 24 h following birth were defined as day 0 of postnatal development (P-0). Within 2 days after birth the offspring was reduced to eight pups. Twenty embryos (ranging in age from day 11 until day 21 p.c.) were used to study the early vasculogenesis. The postnatal development was observed until 30 months after birth (Table 1).

The vascular systems of the brains of 45 animals were filled with a mixture of equal parts of Indian ink and gelatine. After ink injection the brains were fixed in 10% formaldehyde. The brains of 72 rats were fixed by vascular perfusion under a hydrostatic pressure of 120 mmHg for 10 min (fixation solution: 3% glutaraldehyde: analytical grade; 3% p-formaldehyde; 0.1 M cacodylate buffer, pH 7.4, 0.05 M CaCl$_2$). Selected parts of brain slices 1 mm thick were postfixed in a buffered OsO$_4$ solution for 3 h (final concentration: 1% OsO$_4$; 0.1 M cacodylate buffer, pH 7.4), dehydrated in ethanol, and embedded in Epon 812. After this standard preparation technique for

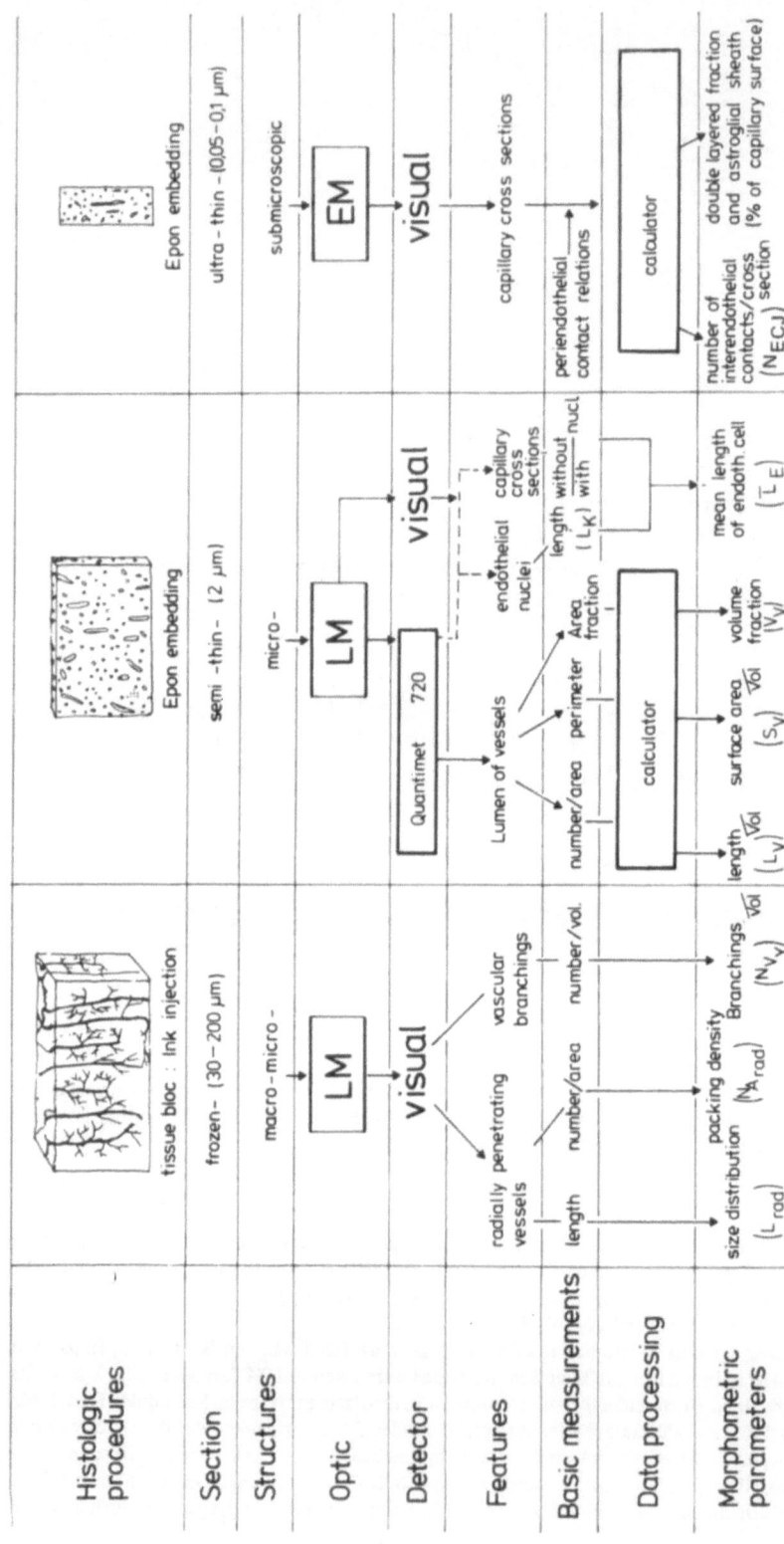

Fig. 2. Outline of investigations which result in the different morphometric parameters

6

electron microscopy the volume changes of brain tissue, measured on single tissue slices, did not exceed 5% (Eins and Wilhelms, 1976). Thirty animals ranging in age from 2 to 70 days p.p. were fixed by vascular perfusion 60 min after an intraperitoneal injection of ^{3}H-thymidine (5-10 μCi/g body weight). Samples of the cerebral cortex were embedded as described above and semithin (2-μm) sections were processed for autoradiography. Serial sections of defined blocs of the occipital cortex (within area 17/18; Fig. 1a) were cut in coronal and tangential section plane.

The section thickness of the ink-injected specimens was 30 μm or 100 μm. The 2-μm Epon sections were stained with a mixture of methylene blue and Azur II according to Richardson et al. (1960). The following data were determined by morphometric methods (Fig. 2).

Length of vertically oriented vascular trunks (L_{rad}). Classification according to the depth of penetration was achieved in 100-μm thick frozen sections of Indian ink-injected material using a linear test system superimposed on micrographs of *vertically* (coronally) sectioned cortex (Fig. 1b). The number of intersections of radially penetrating vascular trunks was counted along ten equidistant lines running parallel to the pia mater.

Volume density of vascular branchings (N_{V_V}). The branchings were counted in frozen sections of 30-μm thickness under the light microscope at a magnification of 200. Thirty fields within lamina IV constituting a total area of 1.44×10^6 μm^2 were examined per brain. The results were not corrected because the overestimation of the number of vascular branchings due to the relation between the size of feature dimension and that of section thickness is arbitrarily compensated by an underestimation due to the fact that branchings that are oriented vertically in respect to the section plane cannot be detected.

Density of vertically oriented vascular trunks (N_{Arad}). The packing density of radially penetrating stem vessels was counted in tangential sections and in frontal sections of ink-injected brains (section thickness, 100 μm). 2-μm Epon sections in tangential plane were used additionally for counting cross sections of vascular trunks in different cortical depths nearly tangential to the pial surface.

[3]*H-Thymidine labeling of endothelial and glia cells.* (Dose and post injection interval, see above) Autoradiograms were prepared by the dipping method (Ilford K2 emulsion) using 2-μm sections of Epon-embedded material.

Mean length of endothelial nuclei (L_K). The projected length of endothelial nuclei along the vessel axis was measured at a magnification of 400 using a calibrated ocular piece. The results are based upon 2-μm Epon, 30-μm frozen, and 8-μm paraffin sections.

Mean length of endothelial cells (L_E). In systematic samples of *capillary cross sections* (i.e., mean internal diameter $\leqslant$ 7.5 μm) the capillary profiles were divided into two groups: Those showing parts of an endothelial nucleus on the sectioned area (mK) and those having no nucleus (oK) (Fig. 3). The frequency of capillary cross sections without or with endothelial nucleus expressed as ratio (oK/mK) is related to the ratio of the collective length of the cytoplasmatic parts and the nuclear parts of endothelial cells. This simplification is possible because nearly all brain capillaries consist of a unicellular sequence of endothelial cells (Niessing, 1950 and present results), i.e., there is no overlapping of endothelial nuclei along the capillary axis. There are no great differences in the dimensions of endothelial nuclei between arterial and venous segments of capillaries. The length of endothelial cells (L_E) is calculated by using the formula $L_E = L_K (1 + oK/_{mK})$ (Bär and Wolff, 1973). The mean length of the endothelial nuclei is measured as described above. The L_E is an average value and differs from the real one because of the systematic neglect of overlappings and indentations of endothelial cell margins and the statistical leveling of individual and regional differences.

Morphometric parameters of brain capillaries evaluated by quantitative image analysis. The measurements were carried out on semithin 2-μm Epon sections using a Quantimet 720 (Cambridge Instrument Company, Dortmund). The tubular network of nearly cylindrical capillaries can be easily detected after perfusion fixation. A further important condition for stereological studies concerns

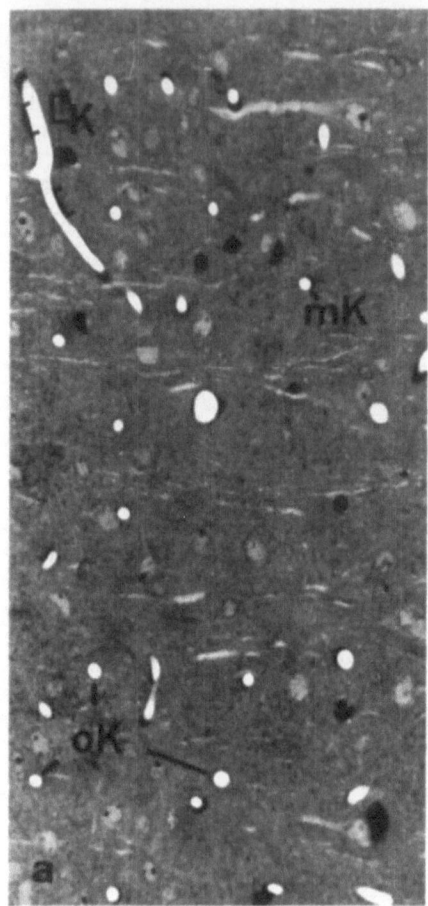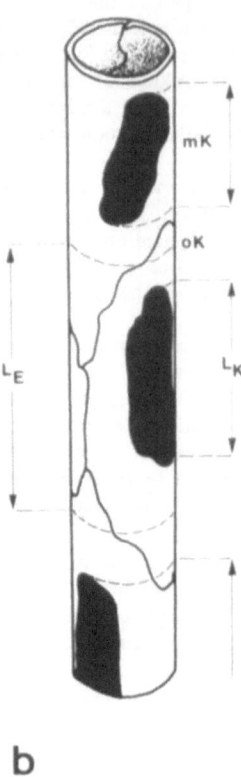

Fig. 3 a. Frontal semithin section (2 μm) through the occipital cortex (lamina IV) of a 6-month-old rat. The lumina of the rinsed vessels are white. x 350; b. Schematic representation of a capillary tube. L_K, length of endothelial nucleus projected onto the axis of the vessel. L_E, Average endothelial length. The mutual indentations and overlappings have been leveled. mK, Part of the capillary tube with nucleus in cross section; oK, part without nucleus in cross section (Bär and Wolff 1973)

the distribution of the tissue element that is to be analyzed. One aim of the present study is to register laminar differences in the capillarization of the cerebral cortex. Therefore a section plane vertical to the pial surface is suitable. Systematic samples were taken in ten measuring fields of equal size (175 μm wide) from pial surface to the border between cortex and subcortical white matter. The localization and the dimensions of these layers differ from the cortical laminae defined by cytoarchitectonic criteria. The orientation of capillary segments in the cerebral cortex was tested by Eins and Bär (1978) who found a random distribution of orientation for cortical vessels having internal diameters < 7.5 μm. This is valid for rats from 14 days after birth onward.

The measuring program runs automatically (Fig. 4). At the beginning data about the material, the experimental conditions, and the thickness of the cerebral cortex must be fed into the calculator, which establishes the optimal microscopic magnification (using Zeiss Planapo 40/ 1.0 oil immersion lens). After adjusting the start position within the section the data input goes on automatically in ten steps (i.e., ten layers) running perpendicularly through the cortex. Four such measuring passes were made in nearly contiguous columns from the pial surface to the subcortical white matter. The four equivalent layers were averaged in each section. The number of

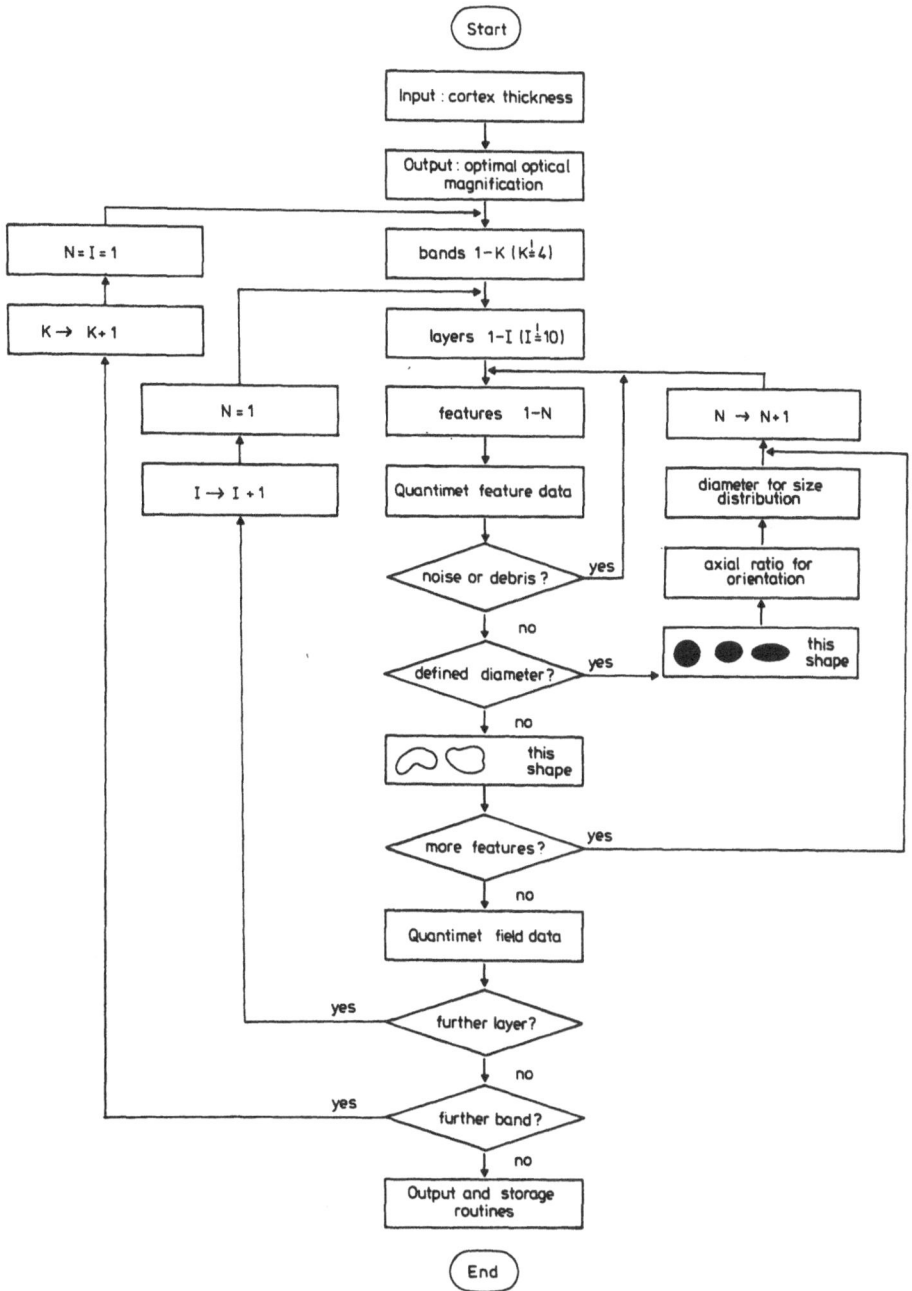

Fig. 4. Flow diagram of the Quantimet programme. Only the single particles (i.e., microvascular profiles) with a definable geometry are used for the determination of diameters. More complicatedly shaped particles are included in the collective stereological data. Each band (K) extends from pial surface to subcortical white matter and is subdivided in ten layers (I, each 175 μm wide). (Original kindly provided by Dr. S. Eins, Göttingen)

vessel sections per measuring field as well as the collective perimeter and the area fraction were measured and transformed into length, surface, and volume of vessels per unit volume of tissue. Additionally, individual data like the area, perimeter, and Ferret diameters were determined for every single section of a capillary. From these data the size classification and the discrimination of noncross sections were calculated. The primary data were processed by a Hewlett-Packard 9830 desk-top calculator combined with a Texas Instruments' Silent 700 printer. The primary measuring errors commonly caused by the influence of section thickness can be neglected because of the optimal relation between the diameter of the microvascular profiles and the section thickness.

Number of interendothelial junctions per capillary cross section. The data are based on random samples of capillary cross sections, i.e., profiles with an internal diameter $\geqslant 7.5$ μm and an axial ratio < 1.5 (magnification x 20,000). The number of interendothelial contact zones (ECJ) per section was counted (Table 2). The relative frequency of vessels having 0, 1, 2, and 3 interendothelial junctional complexes was determined.

Mean thickness of endothelial wall and periendothelial contact relations of brain capillaries. A plastic sheet with ten lines radiating from a central point was superimposed on the intravasal center of random samples of capillary cross sections (magnification x 20,000). Thus, ten points per vascular profile were marked for the analysis of contact relations of the endothelial wall (Fig. 5). The fraction of the basal endothelial surface covered with pericytes and/or processes of endothelial origin were registered on ten intersection points marked by the test lines. In the same way the fraction of capillary surface covered by astroglial processes or lamellae (radial diameter $\leqslant 200$ nm) was estimated.

Oxygen deficiency experiments. Standard laboratory cages were placed in two airtight plexiglass chambers. One chamber was ventilated with an N_2-O_2 mixture of 11% O_2 and 89% N_2. The O_2 concentration was raised to 12% for rats kept in an O_2-deprived environment from days 4 to 14. The chamber containing the cages with the control animals was ventilated with cleaned compressed air (21% O_2, 79% N_2). The circulation in both chambers was adjusted to a flow rate of 5 liters/

Table 2. Number of capillary cross sections used for morphometric evaluations

Age (days after birth)	2/5	7/9	13/15	20	30	180	900
No. of animals	5	5	8	5	2	4	4
No. of electron micrographs	130	230	260	130	200	150	120

Table 3. Number of animals in O_2 deficiency experiments (hypoxia)

Age (days after birth)	Duration of hypoxia (days after birth)	No. of animals	
		Hypoxia (in parentheses, animals labeled with thymidine-H[3])	Controls
14	4– 14	4 (2)	4 (3)
30	14– 30	2 (2)	2 (2)
54/55	4– 14	4	4
	14– 54	3	–
70	30– 70	4 (1)	3 (1)
120	80–120	3	3

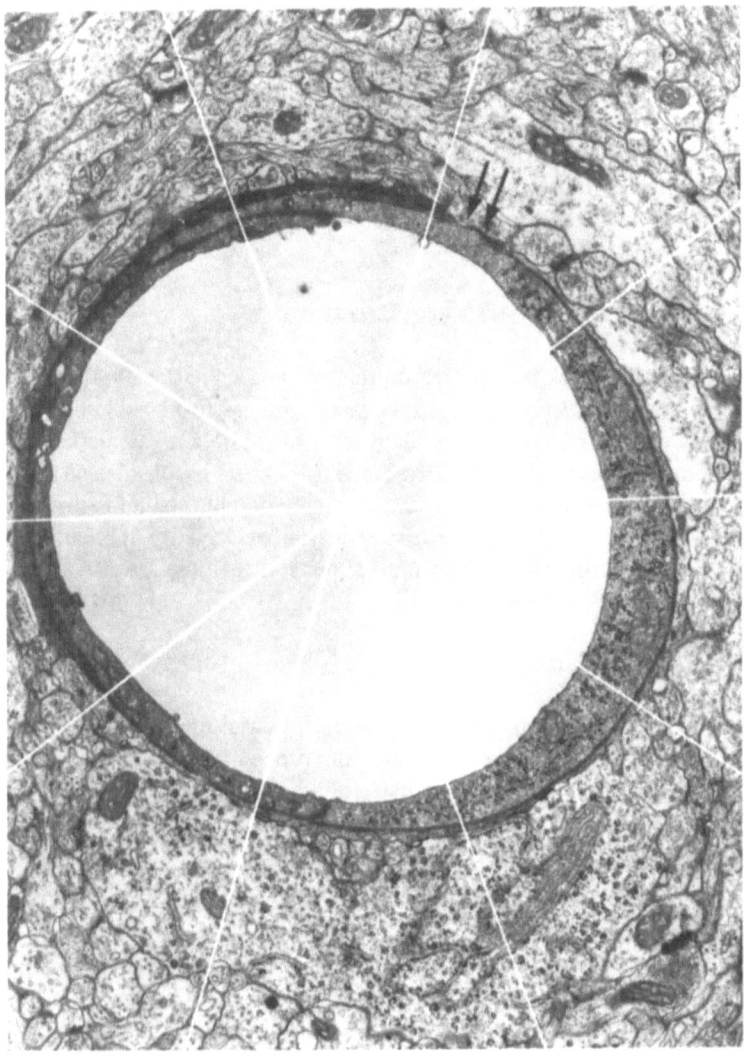

Fig. 5. Capillary cross section from the neocortex of a 14-day-old rat. The screen marks the points in which the perivascular contact relations are examined. The lack of open extracellular gaps and partly incomplete astroglial sheath (*arrows*) are characteristic for the present stage of ontogeny. x 19,500

minute. The CO_2 concentration did not increase over 0.05 vol. % under these circumstances. The temperature in the chambers varied between 22° and 25°C. The relative air humidity ranged between 40% and 85%. It was necessary to open the chambers for a few minutes every day in order to clean the cages and feed and weigh the animals. Shortly after birth the offspring was reduced to six pups (Table 3). Histologic procedures and morphometric techniques were applied as described above.

Statistical methods. Arithmetic mean values of the different age groups were compared and tested (using t – respective F – tests; level of significance, $P < 0.05$. Split plot analyses of variance were made in collaboration with the Institute of Medical Statistics and Documentation, RWTH Aachen.

3 Results

The first cranial vessels, growing from the aortic arch, approach the ventral side of the neural tube and form a *primordial vascular plexus*. The developing hemispheric vesicles are first reached by those vessels at the basolateral surface. As the differentiation of the pallium proceeds, vascular sprouts enter the hemispheric walls perpendicularly. Cerebral vascularization can be divided into an extracerebral and an intracerebral step.

3.1 Extracerebral (Leptomeningeal) Vascularization

Arterial and venous channels that later cover the entire cerebral surface are formed within the *undifferentiated capillary plexus*. In the *perineural* vascular system isolated necroses of capillaries can be observed a few days after intracerebral vascularization has taken place. The regression of the dense *leptomeningeal* vascular plexus starts on the arterial side. In the venous vascular bed a distinct reduction in the number of terminal branchings can be observed during the first postnatal weeks (Fig. 6). Thus, the venous system reaches its definite pattern later in development than the arterial.

3.2 Intracerebral Vascularization

The establishment of the leptomeningeal vascular net is followed by the internal vascularization of the brain tissue. During ontogeny this process extends to different regions of the brain in a characteristic temporal pattern.

3.2.1 First Stage of Intracortical Vascularization

In the cerebral cortex of the rat the development of radially penetrating stem vessels starts 12 days p.c. and is completed during the second postnatal week when the sprouting process of the leptomeningeal vascular system forming new cortical branches comes to an end. The first vascular sprouts enter the pallial anlage before a cortical plate is developed. The branches originating from the radially penetrating stem vessels build up a plexus of undifferentiated capillaries in the ventricular zone situated just opposite the leptomeningeal plexus. Later the intermediate zone of the hemispheric wall which increases in volume is vascularized so that a second capillary plexus can be observed between the cortical plate and the ventricular zone. The terminal arborizations of the radially penetrating vessels are located in the intermediate and in the ventricular zone, whereas the cortical plate lacks capillaries at first. As the cortical plate increases in thickness the vertically orientated vessels must elongate. In the adult cortex the early developed radial stem vessels penetrate all cortical layers almost without side arm branching. In the cortical layers a local gradient of differentiation exists which may parallel the inside-outside layering of cortical neurons. During the tangential growth of the cortical surface the leptomeningeal vascular system develops new branches which enter the cortical tissue. The territories of the terminal branches of these vessels are located in different cortical depths according to their *time of origin* (i.e., old vessels supply the deep layers and younger ones, the superficial layers). To-

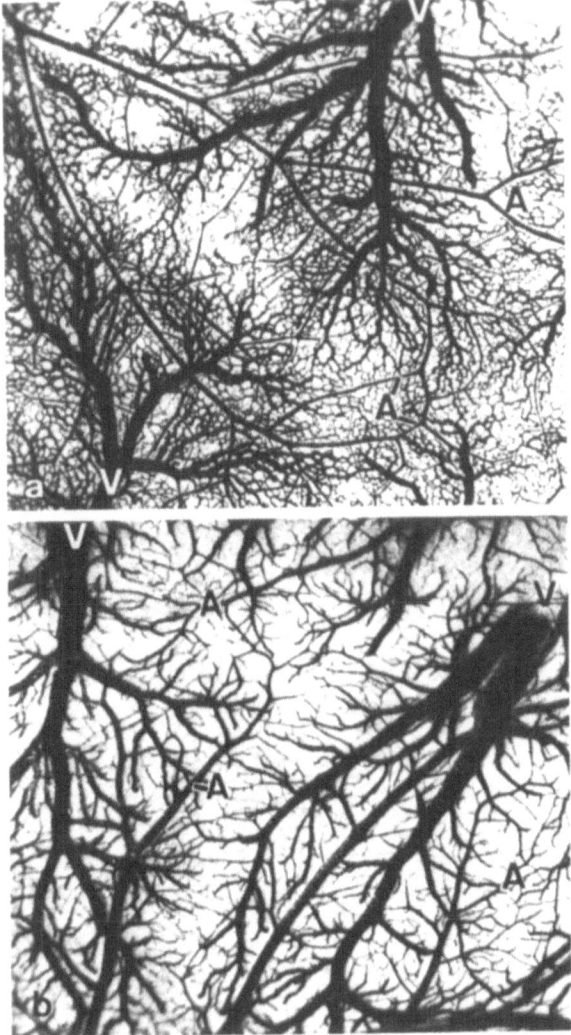

Fig. 6 a, b. Part of leptomeningeal vascular system seen from above (injection with Indian ink). A, arteries; V, Veins. a. 8 days p.p.: dense venous capillary network. x 15; b. 40 days p.p.: the branching density in the venous net has decreased considerably. x 13

gether with the increasing surface area and depth of the cortex the terminal arborizations of the old stem vessels continue to branch and the dimension of the terminal territory of vascular supply enlarges mainly in a tangential direction (Fig. 7). Those vessels that penetrate later in development terminate in younger and more superficially located cortical layers and do not reach the basal zones of the cortex. According to this developmental sequence the oldest vessels are characterized by the longest trunks and the greatest branching territory (Fig. 8). The last and shortest radial vessels grow into the superficial cortical layers during the first week after birth.

After the second week of life *new* vascular sprouts originating from the leptomeningeal system and penetrating the cerebral cortex can no longer be observed. The number of vascular branches that originate from the leptomeningeal system and penetrate the cortex no longer increases. The volume and the surface area of the cerebral cortex continue to increase. Thus, the packing density of the radially penetrating vas-

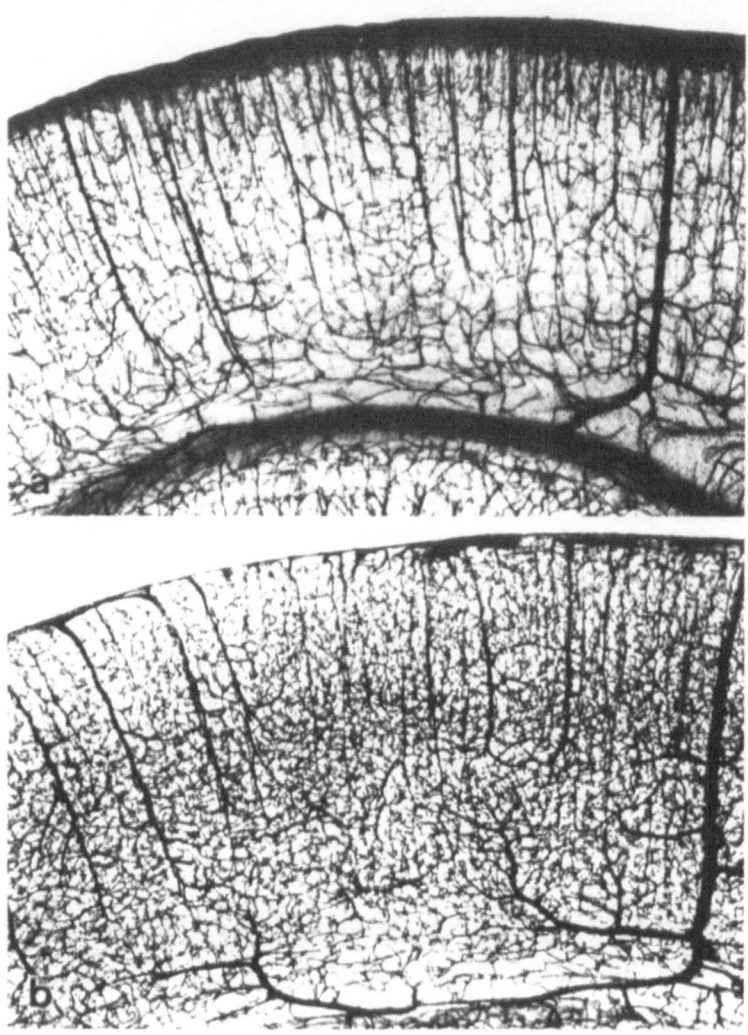

Fig. 7 a, b. Frontal section (0.5 mm thick) through the occipital cortex; 8 days (a) and 40 days (b) p.p. Ink injection, x 41; a. High packing density of radially penetrating vessels, loosely woven, homogeneously distributed capillary net. b. Decrease in packing density and extension of the areas supplied by long radially penetrating vessels; great increase in capillary density, particularly in lamina IV

cular trunks decreases in relation to the further growth of cortical surface area (Fig. 9).

In coronal 100-μm sections of ink-injected material a gradual decrease in the number of radially entering vascular trunks takes place from the cortical surface to the subcortical white matter. In the neocortex three length classes of vascular trunks can be distinguished (Fig. 1 and 10).

The vascular trunks classified according to their length represent a system of hexagonally packed vascular units (modules) of different size (Fig. 11). The smallest modules include the small trunks supplying the superficial cortical layers. Just below, the medium-sized vascular trunks feed and drain the middle cortical zones. The largest vascular trunks are connected to the widest terminal territories in the basal cortical zone

14

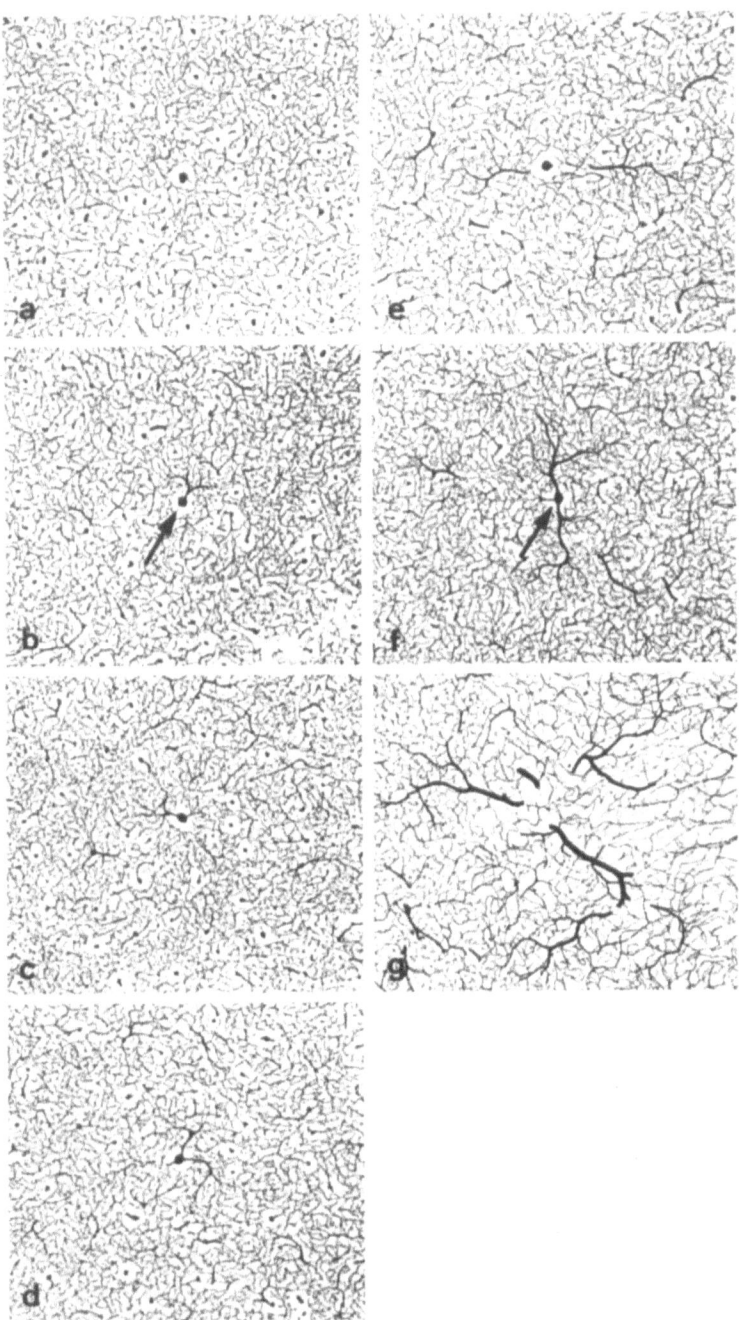

Fig. 8 a–g. Series of horizontal sections (100 μm); starting in the subpial area (*a*) and reaching down to the subcortical white matter (*g*). Vascular system injected with Indian ink (adult rat). x 29. The branching area of the large vessel crossectioned (*arrow*) in the middle of the figures (*a–g*) considerably increases in size in the deep layers of the cortex (*e–g*)

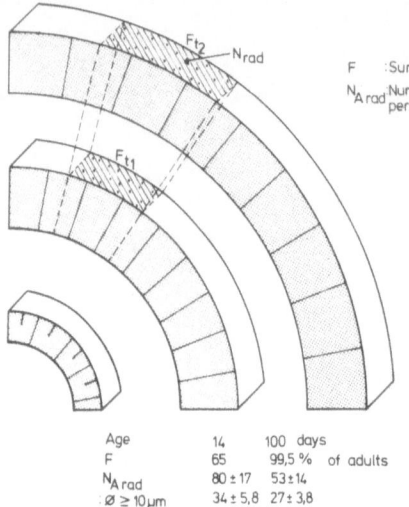

F : Surface area of neocortex

N_{Arad} : Number of vertical vessels per unit test area

Fig. 9. Relations between the increase in cortical surface area and the packing density of vascular trunks. After the late phase of cortical growth (days 14–100) the packing density of the existing radially penetrating vessels (N_{Arad}) has decreased in correlation to the increase in the surface area (from F_{t1} to F_{t2}). During the maturation of the vascular system the diameter histograms of radially penetrating vessels shift to greater size classes

Age	14	100 days
F	65	99,5 % of adults
N_{Arad}	80 ± 17	53 ± 14
$\emptyset \geq 10\,\mu m$	34 ± 5,8	27 ± 3,8

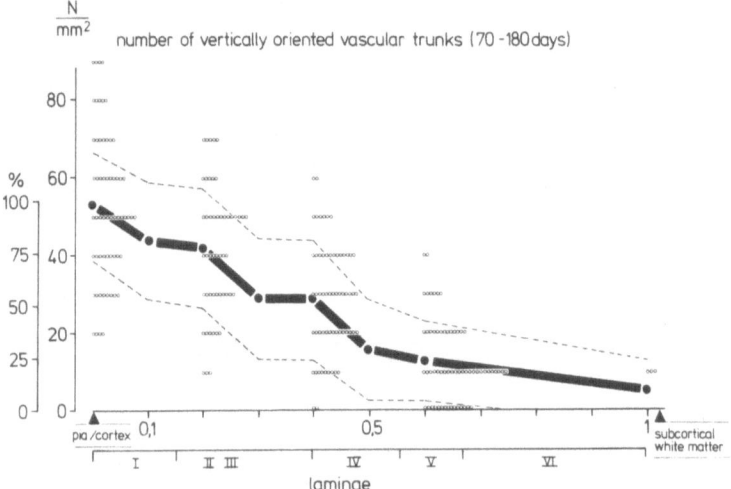

$\frac{N}{mm^2}$

number of vertically oriented vascular trunks (70–180 days)

Fig. 10. Number of vascular trunks in relation to the cortical depth. The *black curve* connects the mean values represented by the *solid dots*. The distribution of single counts is indicated by the *small open circles*. The *broken lines* mark the standard deviation from the mean values

(Fig. 11). In tangential sections just below the pia mater the cross sections of vascular trunks show a typical arrangement according to their caliber (Fig. 12, 13). Vessels with large diameters are surrounded concentrically by vessels belonging to smaller size classes. The distance between central and surrounding vessels increases together with the caliber of the vessels. The largest vascular trunks have the greatest distance from each other. The smaller the vascular trunks are, the more numerous they are and the smaller is their mutual distance. The late and inhomogenous maturation of different cortical areas causes local irregularities of the hexagonal distribution of vascular trunks (Fig. 13).

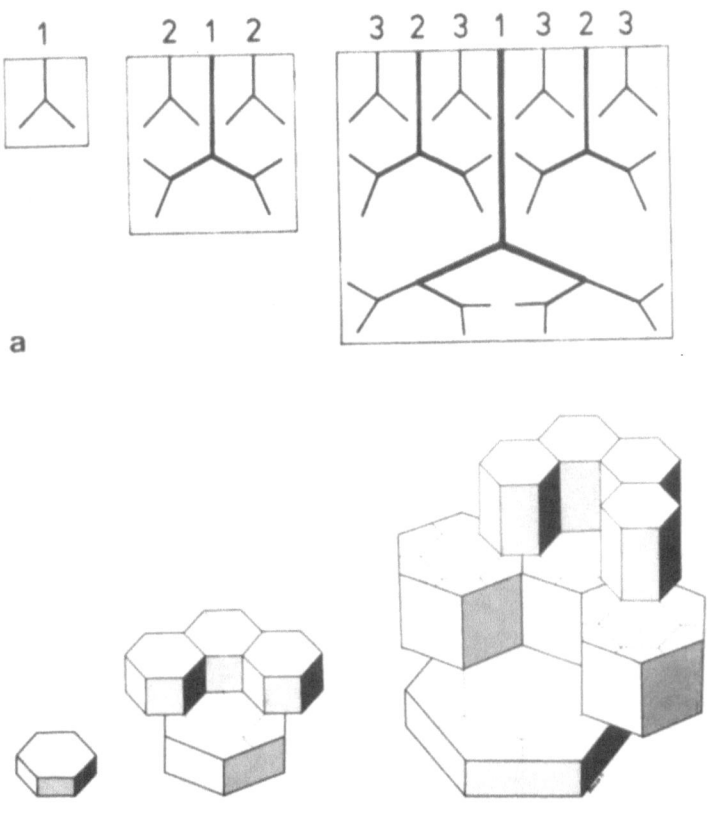

b

Fig. 11 a. Scheme of the successive vascularization of the neocortex during development (after Wolff 1978); b. Three-dimensional model of the arrangement and growth of vascular fields (or modules) supplied by three generations of vascular trunks (*see above*), which had penetrated the neocortex during different ontogenic periods

The vascular trunks are connected by a three-dimensional network of terminal vessels, which is not very dense until the end of the first postnatal week (Fig. 7). The terminal vessels develop into the different segments of the terminal vascular bed. This maturation process involves the endothelium, the basal lamina, and the media of the vessel wall. During embryonic and fetal development the endothelium has irregularly folded surfaces and filopodia projecting into the lumen and between the perivascular cell elements (Figs. 14 and 15). The basal lamina is discontinuous and interrupted by basal processes of the endothelium (Fig. 14). Undifferentiated endothelial cells represent a population of multiplicating cells. New vascular branchings and additional endothelial cells which become intercalated in the existing vascular tube are formed by mitotic division (Fig. 16). Proliferating endothelial cells extend small filopodia into the surrounding tissue (Fig. 15). These filopodia may be responsible for the establishment of intervascular bridges which later become canalized (Fig. 16). The tip of a growing capillary sprout elaborates a growth cone. The possibility of developing new growth cones, i.e. vascular sprouts, a property of all undifferentiated endothelial cells, dis-

17

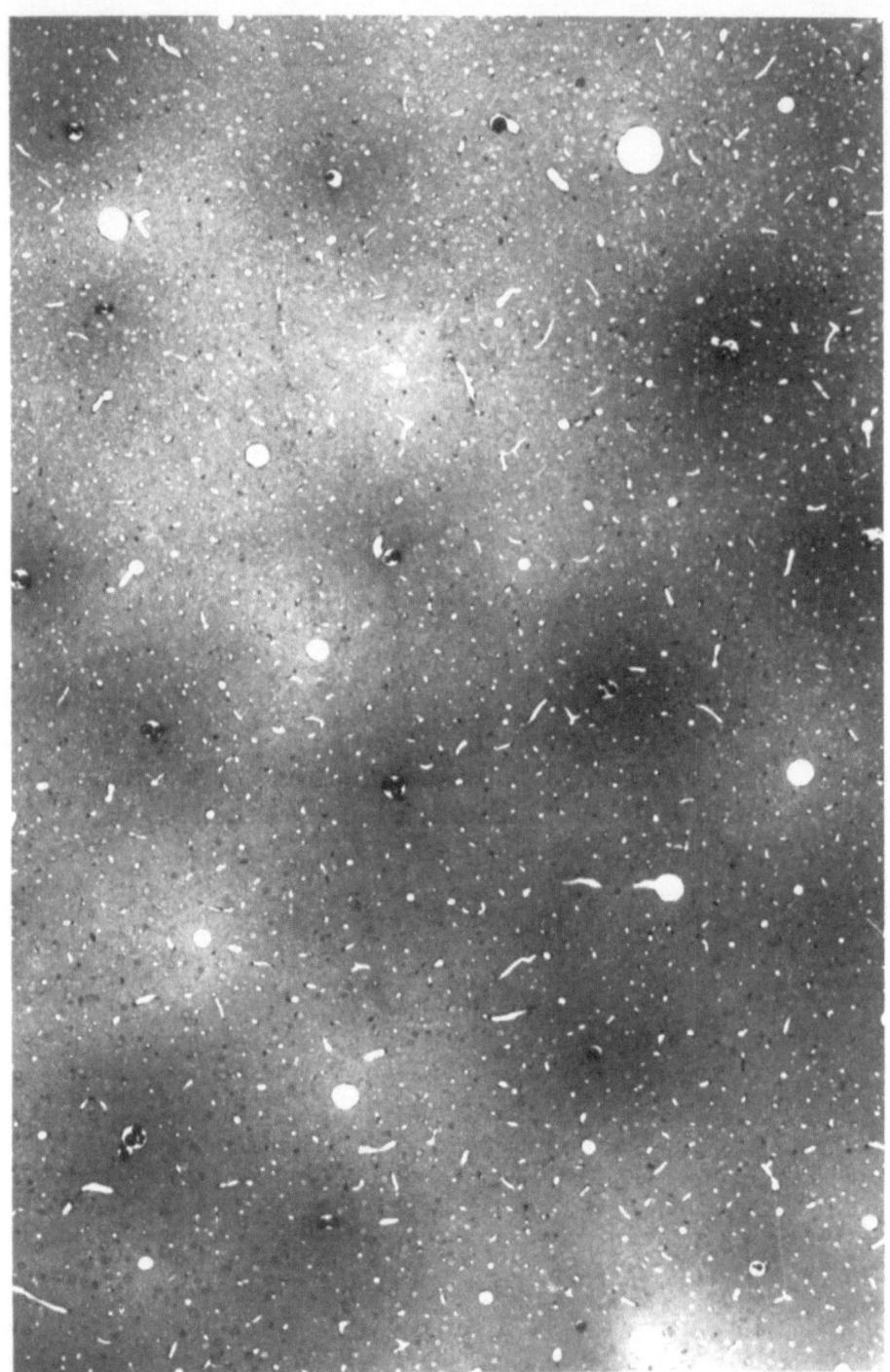

Fig. 12

18

	REL	%	LAM.
•	1	3	W.M.
•	2	6	} V-VI
•	6	17	
·	18	53	II-IV
*	6-7	20	I

Fig. 13. Model of distribution (as seen in subpial, horizontal sections) of the neocortical radially penetrating vessels according to the "system of central places" (Christaller; cited by Schöller 1972). The factual number of the vessels of different caliber in different levels of depth (table on the *left*) corresponds to a hexagonal distribution. However, because of the factual number of small vascular trunks in lamina I (*), disturbances of the hexagonal pattern must be assumed, which may be due to an inhomogeneous regional growth of additional radially penetrating vessels (the original was kindly provided by Prof. J.R. Wolff). *REL*, Numerical relationship between vessels of different size classes (based upon a hexagonal arrangement); %, percent distribution of vascular trunks terminating in different cortical depths (based upon actual countings; see also Fig. 10)

appears as maturation of a muscle layer and a continuous basal lamina proceed within the vascular tree. This maturation process follows a gradient from the proximal (leptomeningeal) parts to the terminal (intracerebral) parts of the cerebral vascular system.

3.2.2 Second Stage of Intracortical Vascularization

During the first stage a basic pattern of vascularization is established which has to adapt to the requirements of the cortical tissue which show *regional-specific differences* later in development. During the second phase of vascularization the density of the terminal vascular bed increases greatly. The increasing capillary surface is associated with the increasing metabolic demand of the cortical tissue. Between days 8 and 20 after birth the endothelial cells of microvessels proliferate rapidly. During this period

Fig. 12. Subpial tangential section through laminae II—III of the occipital cortex. To show the arrangement of radially penetrating arterial and venous trunks, the arteries are shown selectively after embolization with carbonized microspheres (15 ± 5 µm; 3 M Company, St. Paul, Minn.). Arteries and veins are not arranged in pairs. Both vessel types are distributed in a pattern which is more or less hexagonal. x 124

19

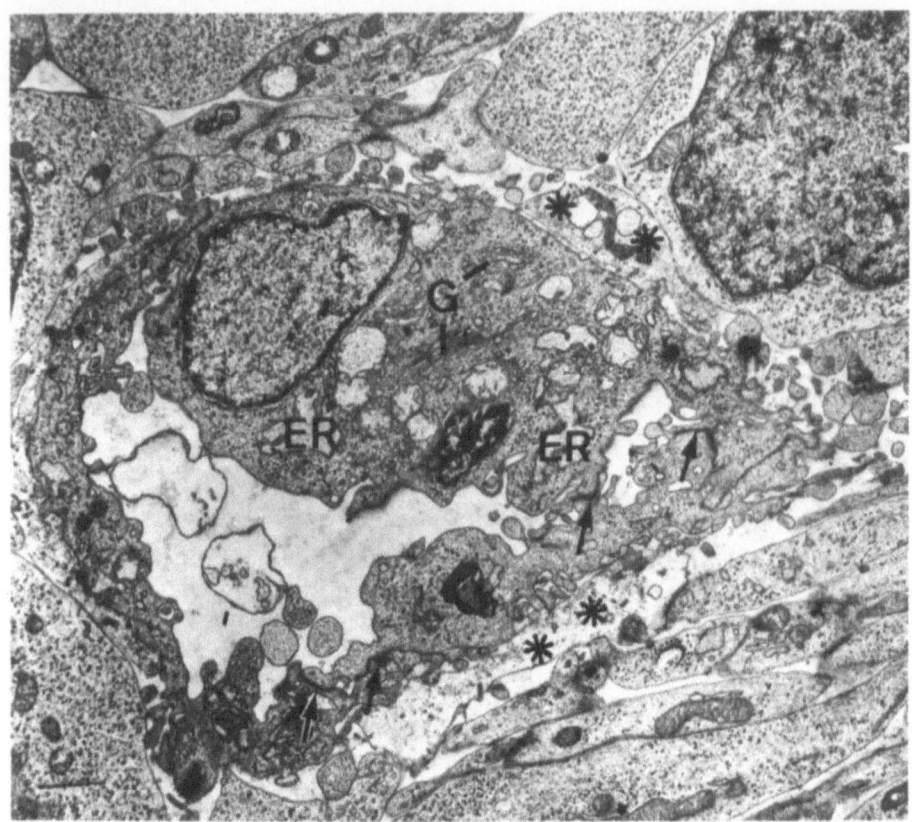

Fig. 14. Immature terminal vessel in cross section (rat neocortex, day 14 p.c., x 9000). The vessel wall is built up by a single layer of several processes of undifferentiated endothelial cells, which are characterized by a volumious perikaryal region bulging into the vessel lumen. Luminal and abluminal plasma membranes of endothelium are irregularly folded. Small cytoplasmic processes project into the lumen and into the adjacent neuronal tissue. Note the well-development Golgi-field (*G*), dilated cisterns of rough endoplasmic reticulum (*ER*), multiple interendothelial junctional complexes (*arrows*). Pericytes are not observed in early stages of capillary formation. Cell processes of astroglial origin locally contact the abluminal surface of the endothelium (*). A basal lamina is not developed. The open pericapillary space communicates with the intercellular spaces between neuronal elements

nearly all vascular branchings are formed. The number of branchings increases from 2000/mm^3 on day 5 to 8000/mm^3 on day 14 (Fig. 17). On day 20 a maximal value of 11,000 branchings/mm^3 is reached. Later the branching density decreases to 9500 in adults. Compared with the changes in branching density the maximal increment of the capillary length per unit volume of cortex ($L_V$mm/mm^3) occurs approximately 3 days later, indicating that newly formed vascular sprouts only influence the collective length of capillaries after the following elongation has taken place (Fig. 17). The L_V in lamina IV increases from 300 mm/mm^3 on day 11 to 1000 mm/mm^3 on day 20 after birth. In contrast to the branching density, L_V continues to increase (Fig. 17; Tables 6 and 7).

During the second stage of vascularization the maturation of the capillary wall proceeds, the endothelial cells of open capillaries are flattened, and the mean length of

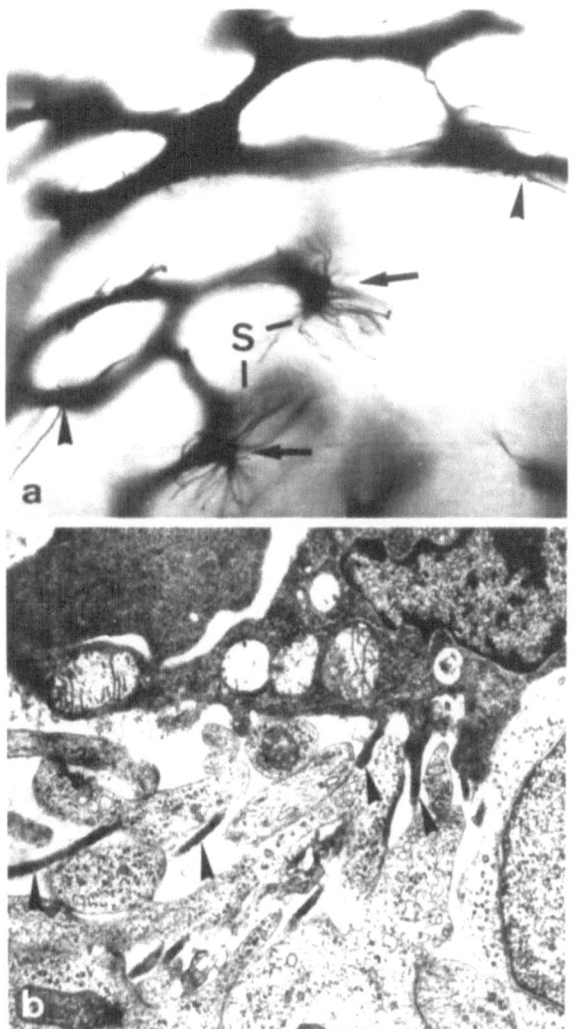

Fig. 15 a. Vascular sprouts (*S*) in the hemispheric wall during the perinatal period. Thin tentacles originate from bulbous capillary growth cones (*arrows*) and/or from preformed immature endothelial cells (*arrowheads*). Golgi-Bubeneite-impregnation, x 490; b. Part of an immature endothelial cell showing thin cytoplasmic processes (*arrowheads*) which project into the neuronal tissue (day 14 p.c.). x 12,300

Table 4. Frequency of interendothelial contact zones (ECJs) in samples of capillary cross sections

Age (days after birth)	Capillary sections (%) with	0	1	2	3	ECJs
2		7.5	30	54	8.5	
5		22	29	46	3	
7		15	13	60	12	
9		15	22	51	12	
13		18	26	45	11	
14		21	18	57	4	
15		22	34	42	2	
20		27	40	27	6	
30		37	37	24	2	
180		32	46	22	–	

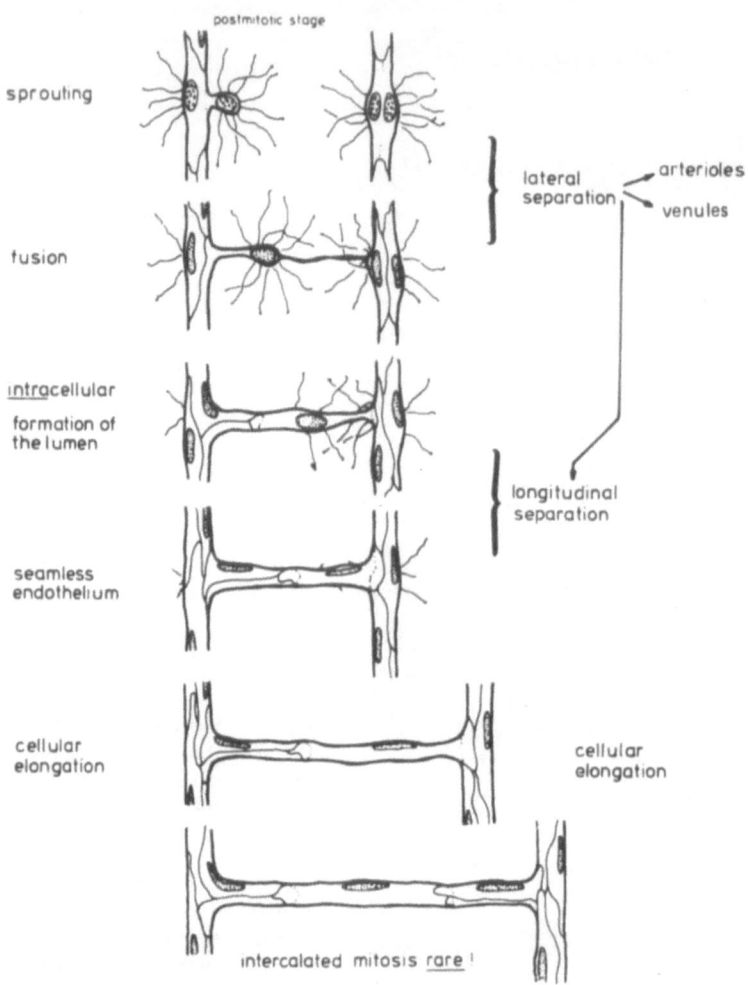

postmitotic stage

sprouting

fusion

lateral separation — arterioles / venules

intracellular formation of the lumen

longitudinal separation

seamless endothelium

cellular elongation

cellular elongation

intercalated mitosis rare !

Fig. 16. Diagram to summarize the formation of net-capillaries in the cerebral cortex (kindly provided by Prof. J.R. Wolff). A fusion of capillary sprouts with preexisting capillaries or with another sprout may be initiated by contact of the small endothelial tentacles. The growth in length of a capillary tube is due to a longitudinal separation of postmitotic endothelial cells followed by cellular elongation. The latter process characterizes the third stage of vascularization. An intercalated mitosis of endothelial cells is a rare event during that period of capillary growth

endothelial cells increases from 26 to 32 μm. The capillary tube consists of a variable number of endothelial cell processes. On the one hand capillary segments built up by several endothelial cells occur; these segments show several interendothelial contact zones on cross sections. Capillaries having only one interendothelial junction along their circumference are the most frequent. On the other hand there are capillary segments that lack an interendothelial contact zone. This form represents intercalated tubular endothelial cells that are perforated by the vascular lumen. During postnatal development the cellular composition of capillary segments changes according to the different frequencies of interendothelial junctions observed on cross sections (Table 4). During the period of rapid proliferation of endothelial cells (first and second post-

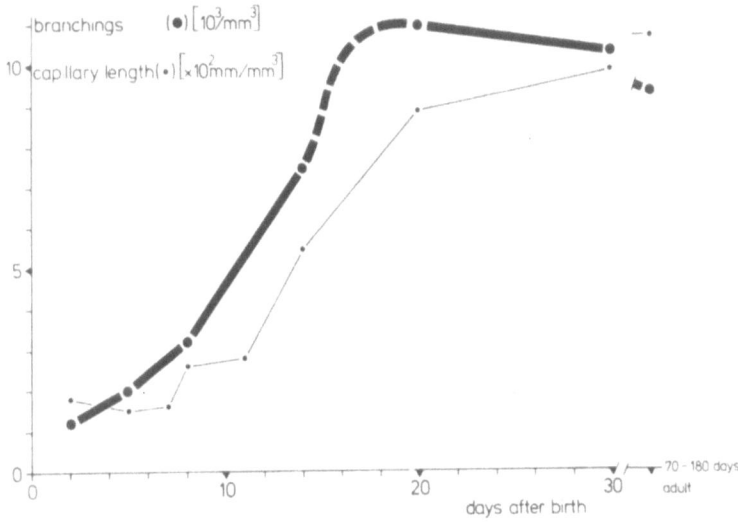

Fig. 17. Numerical density of branchings (*large dots*) and specific length of capillaries (L$_V$, *small dots*) in lamina IV of the parietal cortex during postnatal ontogeny (based on Bär and Wolff 1973). The number of branchings per mm^3 increases considerably during the second and third weeks, and after that decreases to a final value. The increase in L$_V$, however, begins later and continues for a longer time

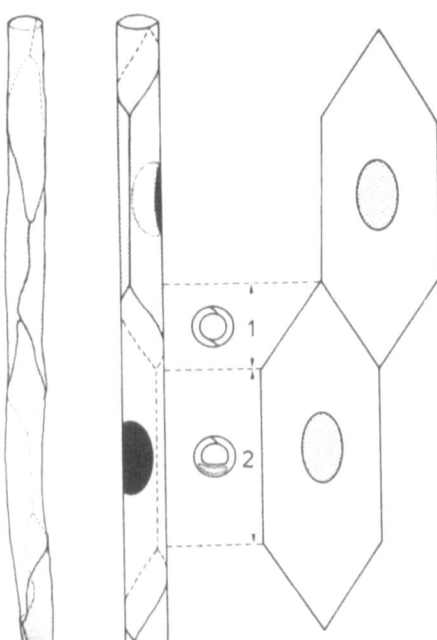

Fig. 18. Schematic view of a capillary tube. The endothelial cells form a row of approximately hexagonal cells that overlap. The ratio in length of those capillary segments that consist of one endothelial cell to those that consist of two cells corresponds to the relative frequency of cross sections with one or two interendothelial contacts (ECJs). *Left*, Arrangement of endothelial cell boundaries after silver impregnation (based upon Lewis, 1933, Fig. 8 a)

23

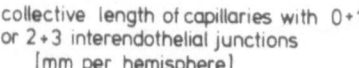

collective length of capillaries with 0+1
or 2+3 interendothelial junctions
[mm per hemisphere]

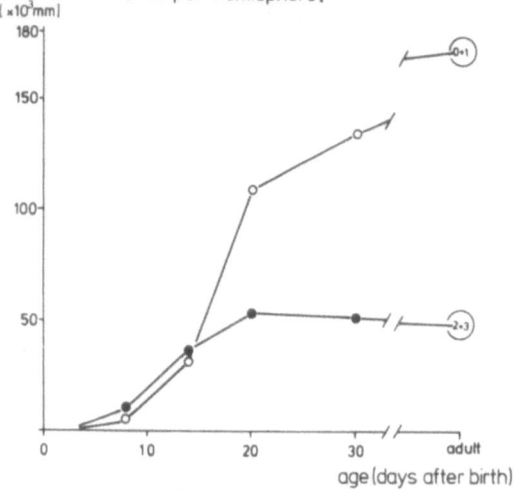

Fig. 19. Absolute length of the capillaries whose walls are composed of one (0 or 1 ECJ) in cross sections or several (2 or 3 ECJs) endothelial cells. From the 20th day onward, only the capillary sections with 0 or 1 ECJ continue to elongate. (Based upon the volume data of neocortex reported by Smith 1934)

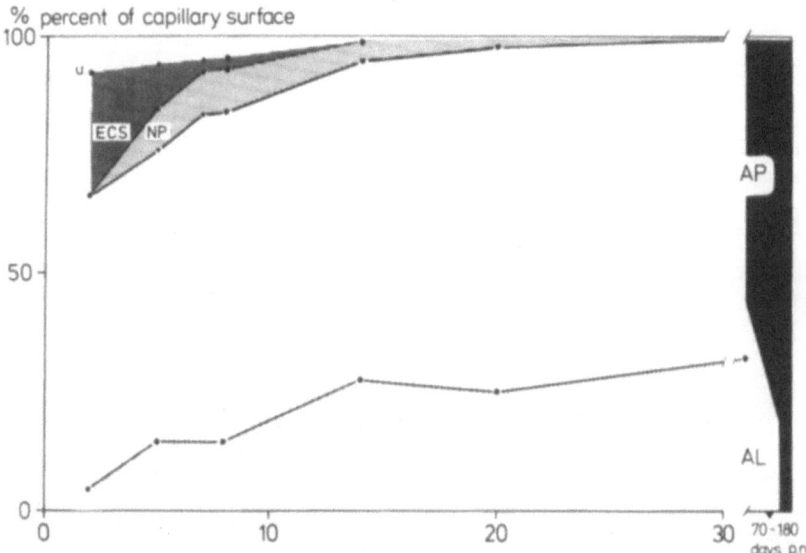

Fig. 20. Postnatal development of pericapillary contact relations. The pericapillary astroglial sheath consists of processes (AP) and lamellar parts (AL, radial diameter ⩽ 200 nm) of astroglial cells. The fraction of AL that covers the vessel wall is presented by the lower curve (dots, mean values in percent of capillary surface; stippled area, range of standard deviation). Hatched region, decreasing extracellular space (ECS) plus nonglial cellular elements (NP) that are in contact with the surface of capillaries. U, Fraction of measuring points associated with nonidentifiable cell processes. Note: The reduction of open extracellular space precedes the completion of the perivascular glial sheath

Table 5. Changes in the capillary wall and in the perivascular glial sheath during postnatal development (in % of capillary circumference). AP main processes; AL lamellar parts of astroglial processes, radial diameter ≤ 2000 Å

Normal Age (days after birth)	2	8	14	20	120
Double-layered capillary wall	51.3 ± 11.5	40.8 ± 4.4	43.4 ± 7.9	29.5 ± 6.8	28.7 ± 5
Perivascular AP	61.4 ± 12	70.6 ± 13.7	63.6 ± 6.8	70.4 ± 11	62.2 ± 12.6
Astroglia AL	4	14.2 ± 9.8	31.4 ± 8.1	25.5 ± 12.6	37.8 ± 12.6
Hypoxia			days 4–14		days 80–120
Double-layered capillary wall			40 ± 15		22.3 ± 7.5
Perivascular AP			60.9 ± 11		56.0 ± 11
Astroglia AL			28.7 ± 10		44 ± 11

natal weeks) capillary cross sections most frequently show two or more interendothelial contact zones. During the third postnatal week the relative frequency of capillary cross sections with two or more contact zones decreases, whereas the number of cross sections with only one contact increases (Table 4). Finally a ratio of 1:2 results, which means that a capillary shows one cell in its cross section alternating with two at the overlapping ends. Thus, the capillary tube is formed by a single file of hexagonal cells around the lumen (Fig. 18).

The fraction of "seamless" capillary cross sections increases during postnatal development. Out of all capillary cross sections 37% are seamless on day 30 (Table 4). Beyond the 20th day of life only the seamless endothelial cells and those parts of capillaries with one interendothelial contact continue increasing in number and/or length (Fig. 19). In immature capillaries pericytes and extensive overlapping of neighboring endothelial cells cause a large fraction of the capillary wall to consist of two or more layers. The extent of the additional layers has been studied quantitatively on capillary cross sections. Without using serial sections one cannot decide in a single electron micrograph whether perivascular processes belong to endothelial cells or to pericytes. Therefore the two possibilities have not been separated. Shortly after birth 50% of the capillary wall is composed of two or more layers. This fraction decreases to 30% on day 20 (Table 5).

In the perinatal period the perivascular investment of astroglia consists of plump cytoplasmatic processes covering approximately 60% of the capillary circumference (Fig. 20, Table 5). During the second and third postnatal weeks slender submicroscopic astroglial lamellae are developed that completely envelop the capillary tube. This is followed by a rapid thickening of the capillary basal lamina. After the maturation of the capillary is completed, sprouting of new branchings was no longer observed.

Table 6. Postnatal development of brain capillaries

Age (days)	Vol. of hemisphere (mm^3)	No. of endothelial cells per hemisphere	No. of endothelial cells per mm^3	No. of branches per hemisphere	No. of branches per mm^3 (Lamina IV)	Length capillaries per hemisphere (mm)	Specific length of capillaries (mm/mm^3)	Length of endothelial cells (μm)
				Relative Increases				
20	170,100	8500	24,000	1,200	4,400	220	24–26	
0– 2	1	1	1	1	1	1	1	1
7– 8	4	4	1	10	2.5	4	1	1
14	6	9	1.5	40	6.5	12	2	1.2
20–21	8	22	2.4	80	10	28	3	1.4
30–35	9	23	2.5	80	9	31	3.4	1.45
55	10	24	2.4	–	–	35	3.5	1.5
Adult	10.5	25	2.4	80	8	38	3.7	1.6
———	210	4.3 x 10^6	20,400	1.9 x 10^6	9,550	169,000	806	40

3.2.3 Third Stage of Intracortical Vascularization

During the last phase of capillary growth, which extends from day 20 until adult hood, the *existing* capillaries continue to elongate (Fig. 17, Tables 6 and 7). At first addit- ional endothelial cells are generated by mitotic division and become intercalated in the capillary tube. The postmitotic cells slide along the vessel axis and become incorporat- ed in the vessel wall. They cause an elongation of the capillary segment. Finally a reduction of the endothelial overlapping results as described above. After pulse label- ing with ^{3}H-thymidine 4%-8% of the endothelial cells are labeled during the second week. In the third postnatal week a sharp drop of mitotic activity of the endothelial cell takes place. In the adult brain the labeling index of endothelial cells decreases to a hundredth of the initial value. L_v, however, increases by about 20% from 670 mm/ mm^3 to 810 mm/mm^3 (Fig. 22). At the same time the mean thickness of the capillary wall decreases from 0.42 (s = ± 0.1) μm on day 14 to 0.26 (s = ± 0.05) μm on day 120 after birth. The mean length of endothelial cells (L_E) increases from 32 μm to 40 μm (Table 7). The relatively small differences in L_E cause an elongation of about 80 mm in a capillary tube which may be composed of 10,000 endothelial cells and distributed in a volume of 1 mm^3. The extent of plasticity of a capillary net is realized by elon- gation and flattening of the existing endothelial cells.

3.3 Morphometric Evaluation of Capillaries in Different Layers of the Cerebral Cortex by Automatic Image Analysis

Capillaries are defined as terminal vessels with an internal diameter of $\geqslant 7.5 \ \mu m$. In the cerebral cortex these microvessels are characterized by a random distribution of orientation. A definition by morphological features of the vessel wall alone is not sufficient.

Size distribution of internal diameters of microvessels. During the first three postnatal weeks there are only minor changes in the mean diameter of microvessels. The distribution of microvascular diameters is always unimodal (Fig. 21). The mode shifts from $6.75 \ \mu m$ on day 8 to $5.25 \ \mu m$ on day 55 after birth. In the early postnatal period the diameters are distributed more uniformly over the observed range than in adult hood, when the histogram curves of the microvascular diameters show steep flanks and a conspicuous peak between 4.5 and 6 μm (Fig. 21). From day 30 until day 690 the distribution curves do not change. The mean internal diameter of capillaries ranges between $5.0 \ \mu m$ and $5.1 \ \mu m$.

Specific length of capillaries (L_v mm/mm^3). During the first postnatal week the mean value of L_v is distributed in a rather uniform pattern and is much lower than later in development (Figs. 22 and 23). Between days 2 and 8 the average value of L_v remains nearly constant. In the second week L_v increases to 417 and in the third week it reaches 670 mm/mm^3 (Figs. 22 and 23). During this period of continuous capillary proliferation and elongation the overall brain growth reaches a plateau. The rapid increment of L_v ends during the fourth week after birth. Subsequently the capillary richness continues to increase but much more slowly. The mean values of single cortical layers show considerable interindividual variations. L_v in different cortical layers varies within the same sample, especially in young animals. Constant inhomogeneities cannot be localized. During the third week the distribution pattern of capillary richness begins to change. The lamina granularis interna forms a conspicuous peak in each curve from 20 days onward (Fig. 23).

When the changes of L_v are analyzed separately in different layers from the pial surface to the subcortical white matter, a developmental pattern is observed which is characteristic for single cortical layers (Fig. 24). During the first 3 weeks the development of L_v is accelerated in *superficial* and *deep* layers, whereas the supply with capillaries is retarded in the *middle* layers. For example 50% of the adult value of L_v is reached 4 days later in the middle zone (5) than in the superficial (1) one (Fig. 24). By 30 days L_v rises to maximal levels in the deep layers, and later it shows an actual decrease. Between days 30 and 55 after birth L_v in layers 9 and 10 exceeds the adult values by about 12% or 6% respectively (Fig. 24). In contrast to this the upper part of the molecular and the deep part of the granular layer show a tendency for L_v to continue to increase slightly for a considerable time.

Volume fraction (V_v%) and specific internal surface area (S_v mm^2/mm^3) of capillaries. If the mean diameter and the mean specific capillary length are known, it is possible to calculate V_v and S_v using equations applied to cylinders provided that the perfused vascular system can be regarded as a three-dimensional net of cylindrical tubes. The independently measured values of V_v and S_v (Fig. 22) and the calculated ones were con-

27

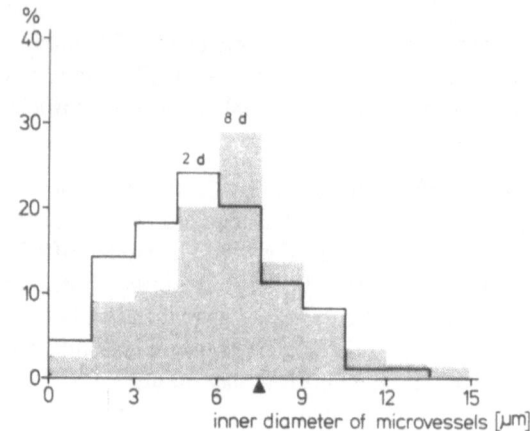

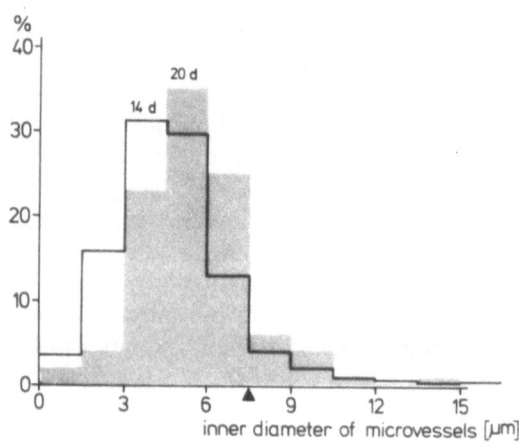

Fig. 21 a–c. Histograms of internal diameters of microvessels from days 2 to 55 p.p. The determination of the mean diameter (D) of capillaries includes all vessels ≤ 7.5 μm in diameter corresponding to the unoriented part of the microvascular system.
a. Day 2: D = 5.0 ± 0.6 μm; day 8: D = 5.2 ± 0.3 μm; b. Day 14: D = 5.4 ± 0.4 μm; day 20: D = 5.3 ± 0.4 μm; c. Day 30: D = 5.0 ± 0.3 μm; day 55: D = 5.0 ± 0.2 μm. (No. of vascular profiles measured: days 2, 8: 360, 680; days 14, 20: 1670, 1100; days 30, 55: 1690, 2550)

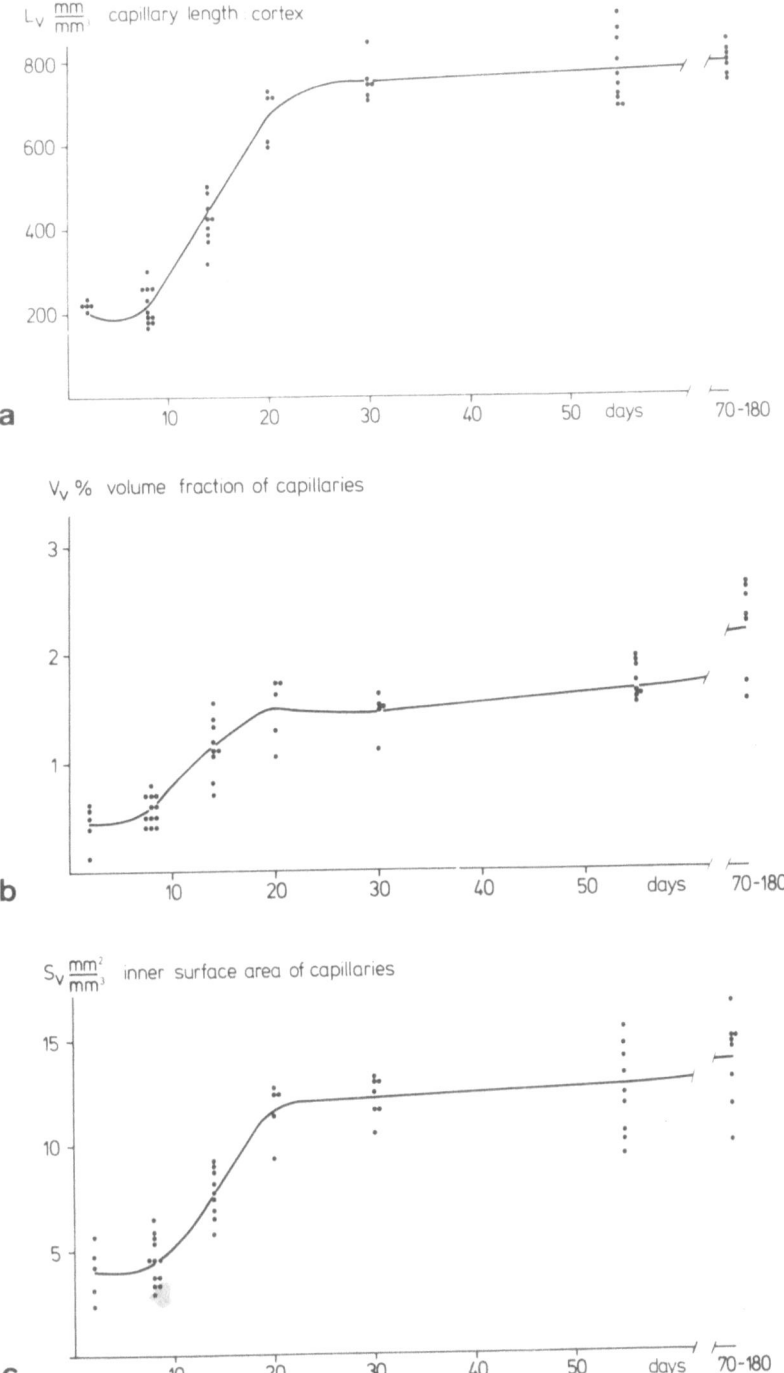

Fig. 22 a–c. Increase in the mean specific length (*a*), in the mean volume fraction (*b*), and in the mean specific internal surface area (*c*) of capillaries in the visual cortex during postnatal development. Each *dot* represents an average value of 40 measuring fields per section. In each section four vertical strips through the cortex were measured. Each strip was subdivided horizontally into ten layers of equal thickness (cf. Sect. 2)

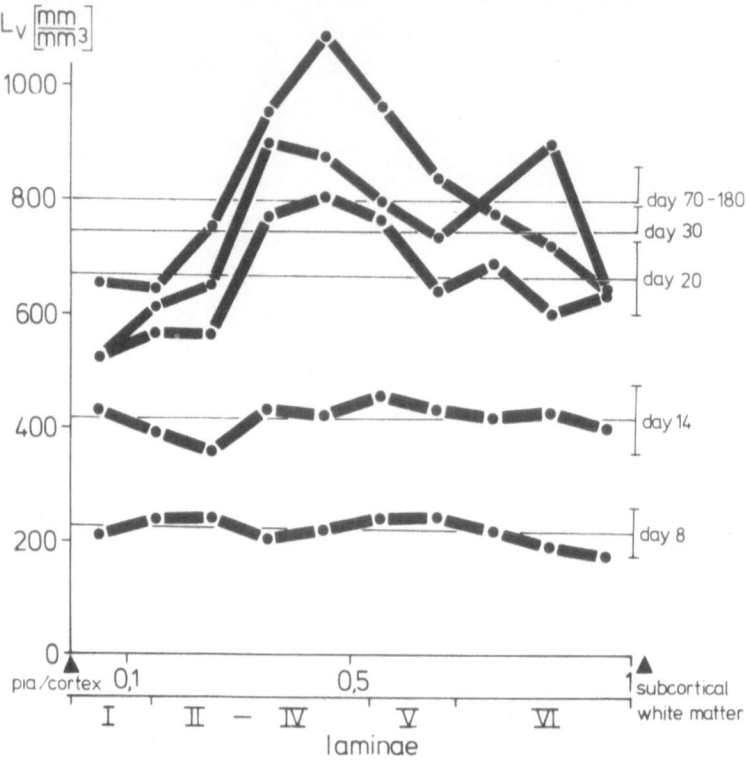

Fig. 23. Postnatal development of L_V in ten layers of cortex from pia mater to the subcortical white matter. The lower scale shows the corresponding cytoarchitectonic laminae (marked with Roman numerals). The overall mean value of layers $1-10$ (± s.d.) is shown for each developmental stage

gruos. Thus the applicability of the tubular model was supported. During development the changes of L_V, S_V, and V_V are interdependent according to the relation between these parameters.

3.4 Thickness of Cerebral Cortex During Postnatal Development

The *absolute* capillary length can only be estimated if the tissue volume and its changes during development are known. In addition to the data concerning postnatal changes of neocortical volume reported by Smith (1934), the thickness of the whole cortex and single cytoarchitectonic laminae were measured in coronal sections through area 17 (Fig. 25). The major growth in thickness occurs during the first 10 days of post-natal development. Cortical thickness increases from 0.6 (s ± 0.05) mm shortly after birth to 1.34 (s ± 0.07) mm (on the 20th day). Within the first postnatal week cortical thickness almost doubles. After the 20th day it decreases to the 10% lower adult value (Fig. 25).

The development of thickness is different for each cytoarchitectonic lamina. The *superficial* (I) and *middle cortical laminae* (II–IV) reach maximal values after 30 days and represent 15% and 43%, respectively, of the total thickness. The prospective *in-*

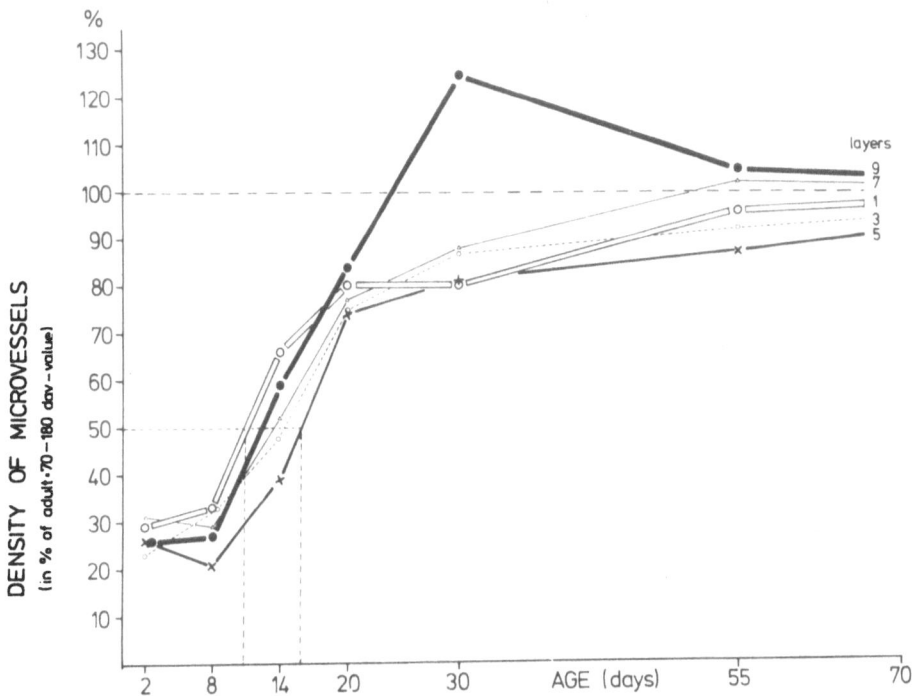

Fig. 24. Postnatal heterochroneous changes in L_V for layers 1, 3, 5, 7, and 9 noted separately. The corresponding mean values of 70 to 180-day-old rats are taken as 100%. In layer 5 the greatest increase in L_V takes place 4 days later than in layer 1. Note the hypervascularization in layer 9 between days 30 and 55

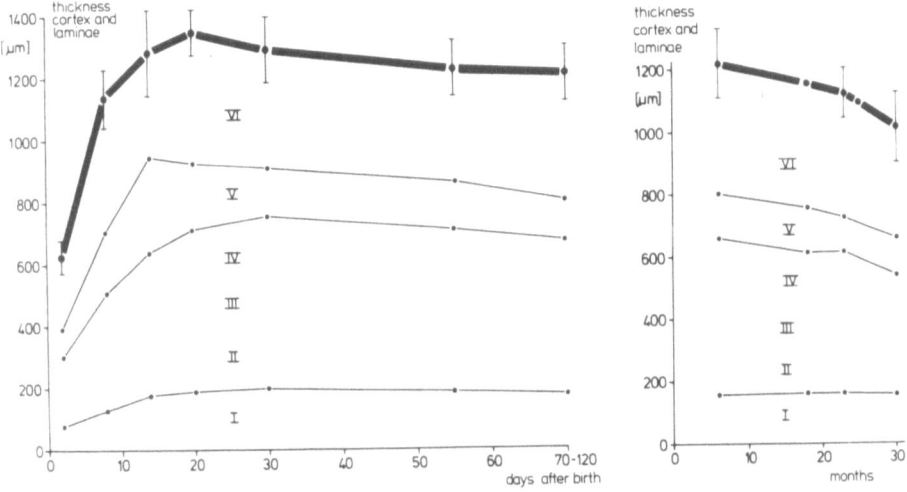

Fig. 25. Changes in the mean thickness of the visual cortex during postnatal life (mean values ± s.d.). The fine drawn curves represent the fraction of the different cytoarchitectonic laminae (I–VI) of the total thickness of the cortex. There are significant ($P < 0.05$, Student t-test) differences between the mean values of 23 and 30 month-old-rats

31

ternal pyramidal layer (V), however, shows an accelerated growth and reaches peak values on day 14 (337 µm, i.e., 28% of cortical thickness). In the adult, lamina V has decreased to 122 µm, i.e., 10% of cortical thickness. The *lamina multiformis* (VI) grows in two steps. After the first week, lamina VI represents 38% of the total cortical depth (according to a mean value of 440 µm). A decrease to 337 µm is observed on day 14. Six months after birth a mean value (410 µm) representing 34% of the total thickness of the visual cortex is reached. With advancing age cortical thickness decreases continuously from 1.22 (s ± 0.1) mm at 6 months to 1.01 (s ± 0.1) mm at 30 months (Fig. 25). The *cytoarchitectonic laminae* are affected by the atrophic process to different degrees. Between 6 and 30 months after birth a reduction of about 25% is detected in the collective thickness of laminae II–IV, whereas laminae V and VI show a decrease of about 14%. The thickness of lamina I remains unchanged during the corresponding period (Fig. 25).

A heterochronic pattern of development and involution characteristic for each cortical area and for single laminae within a given area occurs. The resulting changes in tissue volume can be correlated to the local maturation and to the involution-processes which are characteristic for the single cellular elements of the cerebral cortex.

3.5 Morphometric Evaluation of Capillaries in Different Layers of the Cerebral Cortex by Automatic Image Analysis: Changes During Aging

Size distribution of internal diameters of microvessels. The size distribution curves and the mean diameters of terminal vessels remain almost unchanged between 6 and 23 months (Fig. 26). In single layers (3, 4, 6 and 7), however, the histograms shift to smaller diameters in the visual cortex of 23-month-old rats. In 30-month-old animals the diameter of microvessels, when averaged over the whole cortex, decreases from 5.0 (s = ± 0.3) to 4.4 (s = ± 0.3) µm.

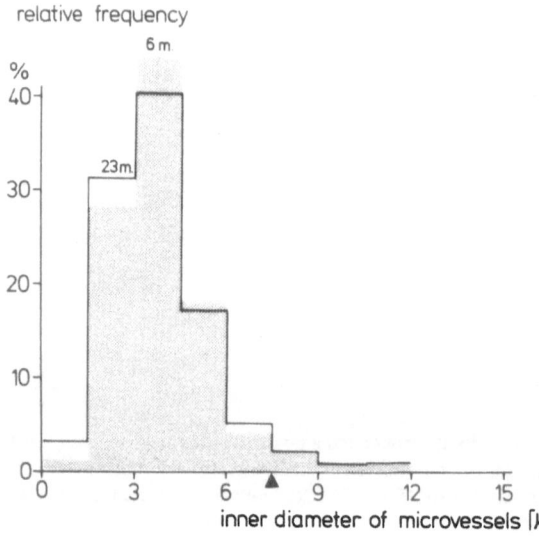

Fig. 26. Histograms of internal diameters of microvessels in the visual cortex of rat. The frequency-distribution curve shifts to lower size classes in the older animal. However, no significant difference can be detected when the general mean values of 6-month-old rats (5.1 ± 0.2) and those of 23-month-olds (4.95 ± 0.3) are compared. Vessels with diameters ≤ 7.5 µm are defined as capillaries

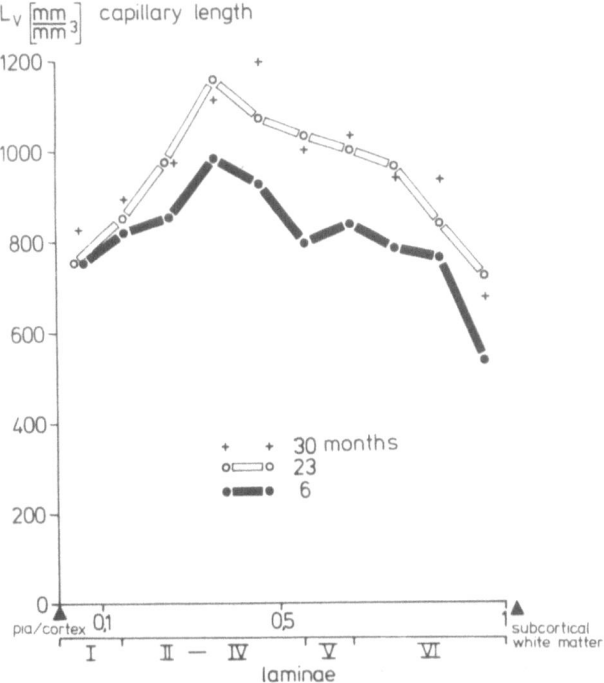

L_v $\left[\dfrac{mm}{mm^3}\right]$ capillary length

Fig. 27. Age changes in the mean length of capillaries per unit tissue volume (L_V) in ten layers from the pia mater to the subcortical white matter. The lower horizontal scale marks the cytoarchitectonic laminae (I–IV). There are significant ($P < 0.05$, Student t-test) differences between the mean values of layer 6 of 6-month-old rats and of 23-month-olds

Specific length of capillaries (L_v mm/mm³). A significant 30% increase of L_v is found within the middle cortical layers of 23- and 30-month-old rats when compared with 6-month-olds. L_v remains nearly unchanged in layers 1 and 2 during this time (Fig. 27).

Volume fraction (V_v %) and specific internal surface area (S_v mm²/mm³) of capillaries. The volume fraction of capillaries decreases during aging in spite of an increasing L_v. In 6-month-old rats V_v amounts to 1.83% (s = ± 0.2), whereas in 30-month-old rats 1.66% of the visual cortex is made up by capillary blood volume. This decrease is caused by the diminution of the mean microvascular diameter. S_v, however, increases from 14.7 mm²/mm³ at 23 months to 15.1 mm²/mm³ at 30 months.

Arithmetic mean length of endothelial cells (L_E). The collective morphometric parameters of capillaries can demonstrate only the overall changes that occur in the microvascular bed during senescence. It is necessary to interpret the observed age changes at the cellular level (Table 7). Therefore L_E and L_K were estimated. The mean length of endothelial nuclei projected into the direction of the vessel axis was found to decrease (6%–8%) in 23- and 30-month-old rats compared to the values in 6-month-olds. In the corresponding period L_E increases from 39 μm to 44 μm (Table 7). In 6-month-old animals a capillary tube of 1-mm length consists of 26 endothelial cells. The cell number per unit length of capillaries is reduced to 23 endothelial cells

Table 7. Postnatal changes of mean length (L_E) and numerical density (N_E) of endothelial cells

Age (days after birth)	L_E (μm)	N_E (N/mm^3 x 10^2)
2	26	85
8	25	89
14	30	137
20	32	207
30	35	212
55	38	206
70	40	200
120	40	200
180	39	209
690	40	233
900	44	217

L_E was estimated by using the equation $L_E = L_K (1 + oK/mK)$. oK/mK, ratio of capillary cross sections without (oK) and with (mK) part of endothelial nucleus in section plane. L_V was used to calculate N_E

at the age of 30 months. The number of capillary endothelial cells per unit tissue volume increases from 20,900 to 21,700 between 6 and 30 months (Table 7). However, the increase in endothelial cell density is not adequate to explain the observed increase in L_v from 809 (s = ± 93) to 962 (s = ± 154) mm/mm^3 which takes place during the same period.

3.6 Effects of Oxygen Deficiency on Intracortical Microvessels

The vascular system grows by *sprouting* (branching) and *elongation* of the endothelial cells. These growth processes are terminated during different developmental periods. The basic pattern of vascular supply is closely correlated to the basic structural organization and to the morphogenesis of the cerebral cortex, whereas the final density and distribution of capillaries depends upon the stage of differentiation and the metabolic demand characteristic for each cortical region and layer. Chronic respiratory hypoxia should cause various changes within the capillary bed according to the developmental period during which it is applied. Additionally the intensity and duration of hypoxia is an important factor influencing the resulting effects. The morphological changes of brain capillaries were checked in 14, 30, 55, and 120-day-old animals after periods of hypoxia which lasted for 10 or 40 days.

3.6.1 Early Postnatal Period (Days 4–14)

The neonatal period is characterized by a rapid increase in the volume of the neocortex. The body weight stops increasing for 3–4 days following the onset of hypoxia. Later it continues to increase as fast as that of the control animals. On day 14 after birth the *mean body weight* of experimental animals is reduced by about 30%. The *mean brain weight* shows a deficit of 11% as compared with the controls. The animals that have survived the period of hypoxia from days 4–14 until day 55 after birth in normal air show mean body and brain weights that are 16% and 9% (respectively) lower than control values. The changes in thickness of the visual cortex were less than 5% after 10 days of hypoxia. After the period of hypoxia the number of 3_H-thymidine labeled endothelial cells exceeds the control level by about 20%-25%. The counts were done in Epon-embedded material after serial sectioning (2 μm). The apparent increase in mitotic activity of endothelial cells induced by hypoxia did not lead to an increased branching density within the capillary net.

Size distribution of internal diameters of microvessels. The respiratory oxygen deficiency is accompanied by a marked dilation of the whole vascular bed (Fig. 28). The histogram curves of microvascular calibers are shifted to greater size classes (Fig. 29). These changes are not homogeneously distributed over the cortical depth. In the upper half of the cortex the mean diameter of capillaries increases from 5.7 to 7.2 μm. (i.e., 25%) after hypoxia from days 4 to 14. In the basal cortical lavers the mean diameter of capillaries exceeds the control values by about 40%. After survival of the experimental (hypoxia) animals until day 55 after birth the dilation of the vascular bed has disappeared. The histogram curves of internal diameters of such animals show a trend to smaller size classes. However, significant differences between the mean values are not found.

Specific length of capillaries (L_v mm/mm^3). Because the experimental conditions cause a strong dilation of the vascular bed it is not appropriate to compare parts of the microvascular beds of experimental (hypoxia) and control animals that are defined by the same diameter range. In order to obtain comparable capillary fractions all microvascular profiles within the range of two size classes on both sides of the median of the size distribution of diameters were counted. This fraction includes about 90% of all vascular profiles. After hypoxia during days 4-14 this "capillary" fraction of the terminal vessels shows an increase of about 10% as compared with the controls. L_v of brain capillaries of animals that have survived until day 55 after hypoxia from days 4 to 14 after birth equals that of the controls of the same age.

Volume fraction (V_v %) and specific internal surface area (S_v mm^2/mm^3) of capillaries. After respiratory hypoxia V_v and S_v increase according to the dilation and the changes of the specific length of the microvascular bed. The volume fraction of capillaries when averaged over the whole cortex amounts to 1.15% (s = ± 0.27) at the age of 14 days. After hypoxia starting on day 4 V_v and S_v have increased to 1.7% (s = ± 0.9) and to 9.2 (s = ± 3) mm^2/mm^3. After hypoxia, higher values of V_v and S_v were found in the upper half of the cortex than in the lower half. Just the opposite is the case in normal controls. The hypoxic dilation of the microvascular bed, however, is usually more pronounced in the deep than in the superficial cortical layers. When animals have survived the period of hypoxia from days 4 to 14 under normal conditions until day 55, the volume fraction of cortical capillaries shows a small but significant decrease from 1.67% (s = ± 0,23) in normal controls to 1.4% (s = ± 0.17) in experimental animals.

Arithmetic mean length (L_E) and thickness of endothelial cells after hypoxia during days 4 to 14. Statistically significant differences in the mean thickness of the endothelial wall were not found after hypoxia in the early postnatal period. The ratio of cytoplasmatic to nuclear parts of endothelial cells had changed, indicating that the

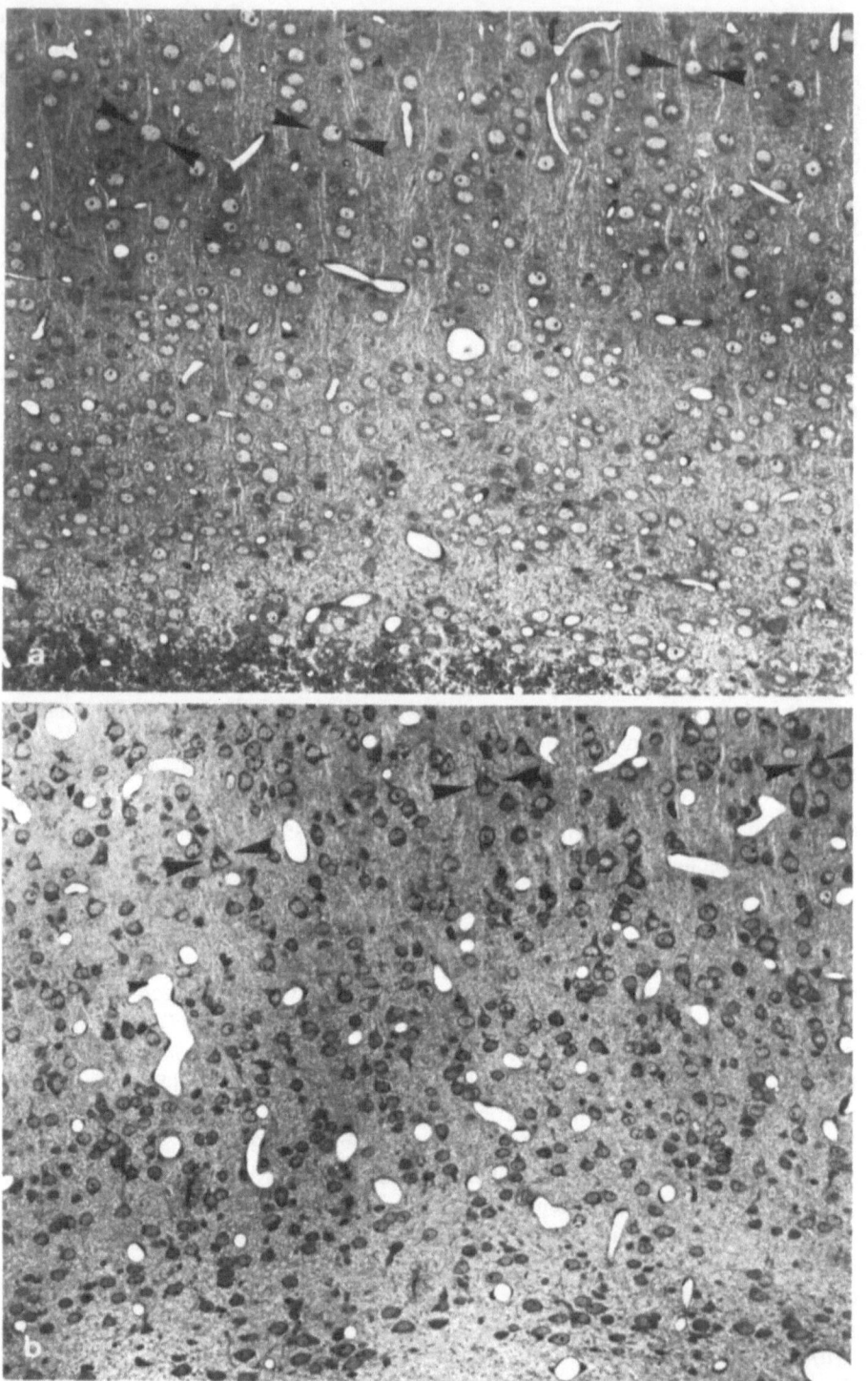

Fig. 28 a, b

relative frequency

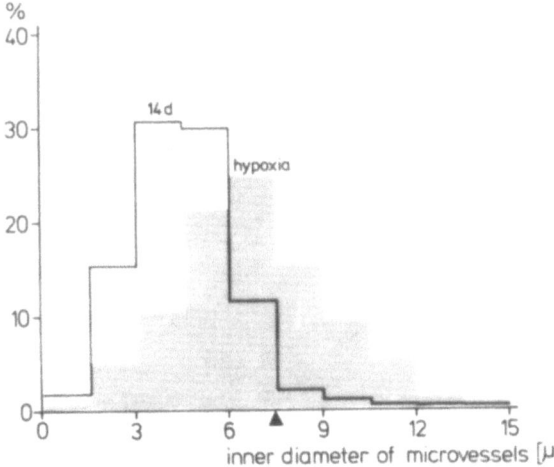

Fig. 29. Size distributions of internal microvascular diameters in the visual cortex of normal rats on day 14 and after oxygen deficiency from days 4 to 14 (*stippled histogram*). Vessels with diameters ≤ 7.5 µm are regarded as capillaries

fraction of endothelial nuclei per capillary length had diminished (i.e., the oK/mK ratio had increased (Table 8). This is caused by an increase in L_E because a corresponding difference in the length of endothelial nuclei projected along the vessel axis could not be found. Thus, the individual size of the endothelial cells increases under conditions of oxygen deficiency. The "double" layered fraction of the capillary circumference was not altered (Table 5).

3.6.2 Days 14 to 54 After Birth

The thickness of the visual cortex is diminished to a mean value of 1.05 mm (control, 1.22 mm) after the experiment (Table 8, Fig. 32).

Size distributions of internal diameters of microvessels. A dilated microvascular bed similar to that already shown in younger animals is observed at the end of the period of hypoxia. The mean internal diameter of microvessels of experimental animals shows a 15% increase as compared with controls (Table 8).

Specific length of capillaries (L_v mm/mm^3). After hypoxia an increase of the mean L_v from 775 (s = ± 82) mm/mm^3 to 1036 (s = ± 51) mm/mm^3 was observed (Figs. 30 and 32). This elevation of L_v is inhomogeneously distributed over the cortical layers (Fig. 30).

Fig. 28 a, b. Lower half of the occipital cortex on day 14 p.p. (coronal semithin sections, x 175). a. Control animal, 14 days p.p.; b. Oxygen deficiency from days 4 to 14: Marked dilation in the terminal vascular bed, considerably retarded myelination in the subcortical white matter, relatively small pyramidal cells in lamina V (*arrowheads*)

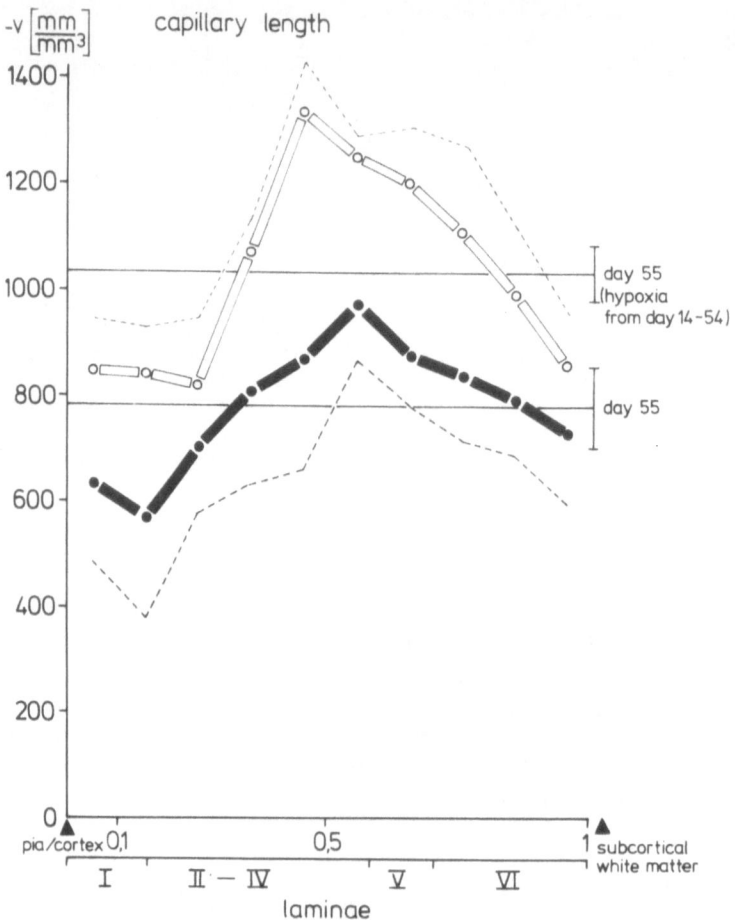

Fig. 30. Changes in L_V in different layers of the visual cortex of the rat after oxygen deficiency from days 14 to 54. The *upper curve* shows deformations which are caused by unequal increases in L_V in different cortical depths. The *dotted lines* represent the standard deviations from the mean values of control (*lower curve*) and of O_2-deprived (*upper curve*) rats. The overall mean values ± s.d. are shown by the *thin horizontal lines*

Volume fraction (V_v %) and specific internal surface area (S_v mm^2/mm^3) of capillaries. After hypoxia V_v and S_v increased from 1.67% (s = ± 0.23) and 12.07 (s = ± 2.31) mm^2/mm^3 to 4.38% (s = ± 1.58) and 22.74 (s= ± 4.64) mm^2/mm^3, respectively.

Arithmetic mean length (L_E) and thickness of endothelial cells. The mean thickness of the endothelial wall decreases from 0.28 μm in controls to 0.22 μm in animals that have adapted to hypoxia (Table 8). An increase of L_E from 38 μm to 40 μm has been calculated based upon the altered relation between the collective length of cytoplasmic and nuclear parts of endothelial cells (expressed as oK/mK ratio).

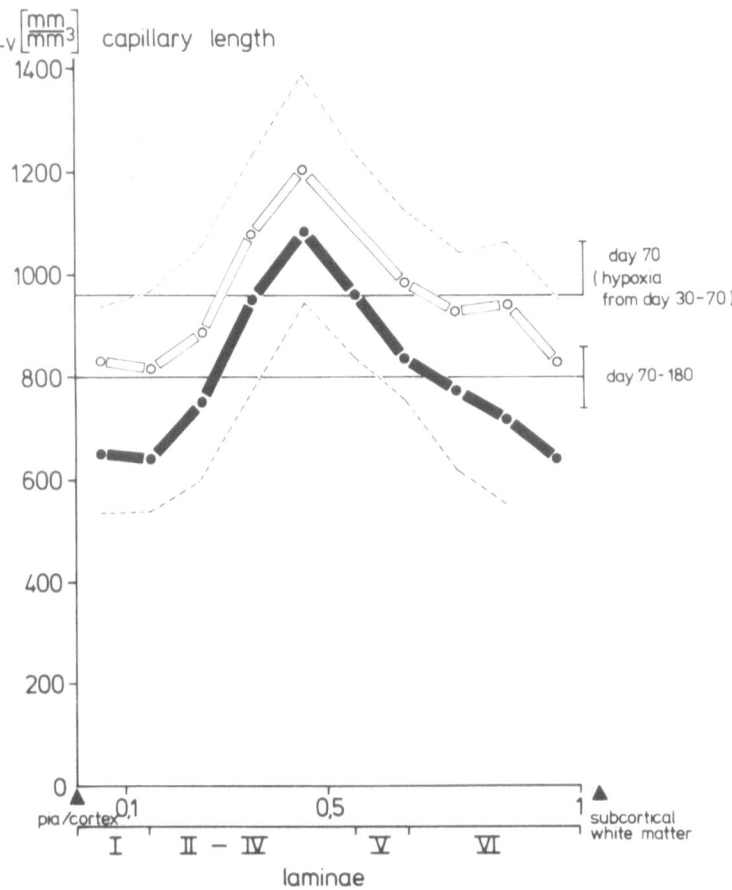

Fig. 31. Specific length of capillaries (L_V) in different layers of the visual cortex of normal 70 to 180-day-old rats and rats subjected to O_2-deficiency from days 30 to 70. Notice the similarity in the general trend of the two curves

3.6.3 Days 30 to 70 After Birth

During this developmental period the neocortex normally reaches its adult thickness. The body weight varies between 300 and 320 g in controls and between 140 and 200 g in experimental rats (Fig. 32). Significant differences between the total brain weights of the two groups were not detected (Table 8). Similarly the thickness of the visual cortex is almost unchanged after hypoxia during days 30 to 70 (Fig. 32).

Size distributions of internal diameters of microvessels. The histogram curves shift to larger size classes of diameters, resulting in an increased mean diameter (6.5 μm after hypoxia, 5.3 μm control value) (Table 8).

Specific length of capillaries (L_v mm/mm^3). An increase of the mean L_v from 760 mm/mm^3 to 960 mm/mm^3 was detected (Fig. 31). L_v is evenly elevated throughout the cortical depth.

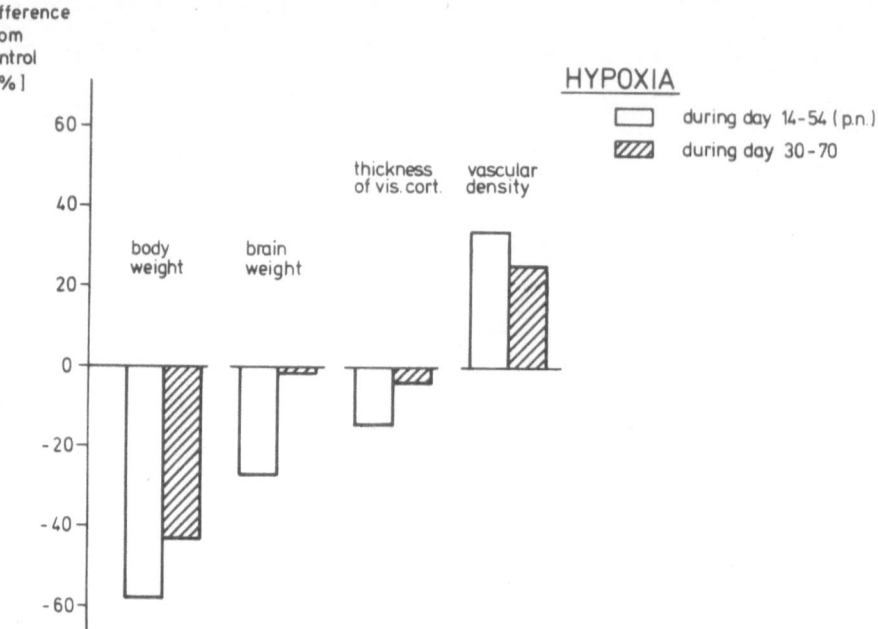

Fig. 32. Survey of changes in various morphological parameters after oxygen deficiency. Oxygen deficiency from days 14 to 54 causes a marked decrease in brain weight and cortical thickness. The same parameters only change insignificantly when equivalent experimental conditions are applied from days 30 to 70

Volume fraction (V_v %) and specific internal surface area (S_v mm^2/mm^3) of capillaries. V_v represents 2.2% (s = ± 0.4) of cortical tissue in controls and 4.8% (s= ± 1.9) in rats adapted to the hypoxia. S_v rises from 13.9 (s= ± 2.1) to 21.8 (s = ± 5) mm^2/mm^3 during the same period.

Arithmetic mean length (L_E) and thickness of endothelial cells. After hypoxia from days 30 to 70 no differences in the thickness of the endothelial wall were observed between hypoxic and control rats. Dramatic changes in the endothelial cells were absent, only a small increase in L_E to 41 μm compared with 40 μm in controls was detected.

Periendothelial contact relations. The double layered fraction of endothelial surface is diminished to 20% in the experimental animals compared with the control value of 30%. The pericapillary astroglial processes or lamellae completely surround the capillary surface. Under hypoxic conditions a relative expansion of astroglial lamellae occurs.

3.6.4 Days 80 to 120 After Birth

Hypoxia during this period did not result in changes of cortical thickness.

Size distributions of inner diameters of microvessels. The histogram curves shift to larger size classes after hypoxia, resulting in a 25% increase of the mean diameter (Table 8).
Control 5.3 (s = ± 0.2) μm; Hypoxia 6.6 (s = ± 0.16) μm

Table 8. Effects of oxygen deficiency on cortical thickness and on intracortical capillaries

Age (days after birth)	Mean brain weight (g)	Mean thickness of visual cortex (mm)	L_V (mm/mm³)	Mean internal diameter of microvessels (µm)	Capillary cross sections (oK/mK)	Thickness of endothelial[b] wall (µm)
14	1.05	1.19 (± 0.14)	428 (± 43) $n = 9$	5.7 (± 0.7) $n = 2320$	1.76 (± 0.36) $n = 1239$	0.42 (± 0.1) $n = 32$
4–14 Oxygen deficiency	0.91	1.13 (± 0.08)	460 (± 40) $n = 9$	7.7 (± 1.1)[a] $n = 2120$	2.06 (± 0.6) $n = 771$	0.41 ((± 0.08) $n = 30$
55	1.83	1.22 (± 0.09)	775 (± 82) $n = 10$	5.6 (± 0.1) $n = 1800$	2.60 (± 0.8) $n = 1753$	0.28 (± 0.06) $n = 11$
14–54 Oxygen deficiency	1.51	1.05 (± 0.04)[a]	1036 (± 51)[a] $n = 4$	6.5 (± 1.0) $n = 1900$	2.90 (± 0.76) $n = 1227$	0.22 (± 0.04)[a] $n = 10$
70	1.90	1.38 (± 0.09)	760 (± 66) $n = 5$	5.3 (± 0.46) $n = 2100$	2.50 (± 0.58) $n = 1214$	0.24 (± 0.05) $n = 10$
30–70 Oxygen deficiency	1.80	1.35 (± 0.08)	960 (± 104)[a] $n = 10$	6.5 (± 0.86) $n = 6000$	2.67 (± 0.6) $n = 1281$	0.24 (± 0.04) $n = 15$
120	1.91	1.21 (± 0.09)	832 (± 34) $n = 4$	5.3 (± 0.2) $n = 1700$	2.50 (± 0.57) $n = 1826$	0.26 (± 0.05) $n = 21$
80–120 Oxygen deficiency	1.82	1.18 (± 0.04)	875 (± 115) $n = 4$	6.6 (± 0.16)[a] $n = 2000$	2.70 (± 0.64) $n = 2406$	0.22 (± 0.06) $n = 30$

Values represent means (± s.d.): [a] significantly ($P < 0.05$; t-test) different from controls; [b] the nucleated parts of the endothelial wall were deleted

Specific length of capillaries (L_V mm/mm³). The comparison of the mean values of L_V of control vs experimental rats did not result in significant differences. The mean values of L_V when averaged over the whole cortex are 832 (s = ± 34) mm/mm³ in controls and 875 (s = ± 115) mm/mm³ in experimental rats.

Arithmetic mean length (L_E) and thickness of endothelial cells. The mean thickness of capillary endothelial cells ranged from 0.26 (s = ± 0.05) µm in controls to 0.22 (s = ± 0.06) µm in experimental animals.

L_E did not vary. However, L_K decreased from 11.5 (s = ± 1.6) to 10.6 (s = ± 1.4) µm under hypoxic conditions.

Periendothelial contact relations. The increase in capillary diameter caused by the hypoxic dilation is not accompanied by an adequate extension of pericytic processes. The relative fraction of endothelial surface covered with pericytic and/or periendothelial processes decreases by about 20%. The perivascular astroglial sheath becomes flattened after the period of hypoxia (lamellar parts of perivascular astroglia, 37.8% in controls, 44% in hypoxic rats) (Table 5).

4 Discussion

4.1. Methodological Considerations

Arterial and venous trunks penetrate the cerebral cortex perpendicularly and terminate a different levels of depth. These stem vessels are connected by a three-dimensional network of capillaries (Pfeifer 1928) that lack a preferential orientation (Eins and Bär 1978). To characterize a capillary net by morphometric means, parameters like the collective length of vessels, the vascular diameter and surface area, and the number, position, and angles of branchings should be estimated. In the present study, the counting of vascular branchings must be omitted in the routine measurements because this feature is not unequivocally recognized by television image analysis. The *number* and *position* of branchings in a microvascular net, however, are very important for the perfusion properties and the oxygen supply of the surrounding tissue (Grunewald 1969). Without considering the branchings of a capillary net the functional interpretation of morphometric data is limited.

Compared with other components of the cortical tissue the microvascular system represents optimal prerequisites for television image analysis. After a standardized perfusion fixation, the preparative volume changes of cortical tissue do not exceed 6% (Weibel 1969, Eins and Wilhelms 1976). Therefore a correction of the estimated morphometric values is not necessary. The empty vascular lumen can be detected selectively by automatic image analysis without time-consuming interactive techniques, e.g., using a light pen. Other histologic techniques may be applied, but there are several disadvantages. When using methods like intravasal staining of erythrocytes or filling the vascular system with colored gelatine mixtures, the completeness can hardly be verified, and estimation of the vascular diameter is difficult because of the variable filling of the vessels. Histochemical methods (e.g., alkaline phosphatase reaction) show variations of intensity and of distribution of the reaction product, which depend upon the age and the experimental conditions (Källén and Valmin 1960; Friede 1966; Müller et al. 1971; Woossmann et al. 1977).

Several authors have assumed that brain capillaries do not possess a preference of orientation (Diemer 1963; Burian 1970; Saunders and Bell 1971; Wiederhold et al. 1976). Recently, random distribution of orientation was proved for cortical capillaries (i.e., internal diameter < 7.5. μm) of rats more than 8 days old (Eins and Bär 1978). Thus, a stereological analysis of cortical capillaries can be made by measuring microvascular profiles sectioned in a fixed (e.g., coronal) cutting plane, which is appropriate to detect layer-dependent differences in capillarization. An important factor that may influence the resulting morphometric parameters (L_v, S_v, V_v) after perfusion fixation is the potential existence of unperfused capillary segments. A reserve of closed capillaries which open under different physiologic and experimental conditions as described in other organs was suggested to exist in the brain too (Cobb and Talbott 1927; Opitz and Schneider 1950; Weiss and Edelman 1976). In stained 2-μm Epon sections the completeness of perfusion can be checked. Unperfused or closed parts of capillaries represent less than 1% after standard perfusion fixation.

The dynamic changes in the capillary system in the adult brain seem to be restricted mainly to variations of the vessel diameter, whereas the capillary length shows only minor changes under different pathologic (Hirsch and Schneider 1968) and experimental conditions (Bär et al. 1975). After a period of aerogenic hypoxia from day 80 until day 120, L_v did not differ from control values. These results do not agree with a suggested opening of previously closed capillaries accompanying the increase in regional cerebral blood flow after hypoxia (Weiss and Edelman 1976). Thus, changes in the regional cerebral blood flow seem to be due solely to variations of the diameter of resistance vessels (Kontos et al. 1978). Disturbances that exceed the regulatory capacity of the vascular system may cause damage of nerve cells before further vascular adaptation by proliferation of endothelial cells is realized. Under conditions of increased metabolic demand (e.g., drug-induced seizures) the adequate blood supply of the brain tissue seems to be limited by the existing vascular system.

4.2 First Stage of Intracortical Vascularization and Architecture of Intracortical Vascular Trunks

The early intracerebral vascularization is characterized by the formation of *new sprouts* originating from endothelial cells of undifferentiated segments of the leptomeningeal vascular system. In the leptomeningeal system, the development of new capillary branches is completed during the second week after birth, coinciding in time with the morphological maturation of the vascular wall. The leptomeningeal branches penetrate the cortical tissue perpendicularly and grow toward the subventricular zone, where a capillary plexus is built up opposite to the leptomeningeal vascular system (Strong 1964). A similar vascular pattern is observed in many parts of the neural tube and seems to be associated with the development of *laminated neural structures* (Feeney and Watterson, 1946; Ismailowa, 1958; Strong, 1964; Lazorthes et al., 1968; Menkes et al., 1975; Wolff, 1976 a, b, 1978). A constant chronological sequence and topographical pattern of the earliest neural vessels are reported (Spalteholz, 1923; Feeney and Watterson, 1946; Strong, 1961; Camosso et al. 1976), but the induction of vascularization, whether it depends upon the increasing distances between inner and outer surfaces of the neural tube or whether chemical factors are responsible, is unknown as yet.

Between extra- and intracerebral vascular plexus a *metabolic gradient* is built up which may influence the distribution of cells within the hemispheric wall. Along this gradient a layering of neurons might take place according to their metabolic demands. After hypoxia a selective damage of cells occurs in distinct laminae of the optic tectum (Menkes et al., 1975). This observation suggests that a gradient of oxygen may play a role during the morphogenesis of laminated structures. The morphogenesis and cytodifferentiation of the neuroepithelial cells may be influenced by the fraction of intracortical vascular (i.e., mesodermal) surface as observed in other tissues (Masters 1976). The mesodermal (meningeal or vascular) contact of a neuroepithelial cell is an important step toward the determination of the neuronal or glial cell line (Rickmann and Wolff 1976). Thus, an interdependence exists between the initial structural organization of the cerebral cortex and the distribution and surface area of intracortical vessels.

During morphogenesis a characteristic angioarchitectonic feature of the cerebrovascular system is established. The main branches of the large extracortical leptomeningeal arteries do not send out intracortical side arms (Kapustina 1952). The intracortical branches have their source in more distal parts of the leptomeningeal vascular system. The more distal (terminal) an intracortical branch originates from the leptomeningeal net, the smaller is its caliber (Kapustina 1952) and its length.

Thus, a correlation exists between the caliber size of the main vessel and the size of the branching one. This feature may reflect a developmental sequence of the intracortical arteries as suggested by Wolff (1976a) who assumed an interdependence between the length of cortical arteries and their time of origin. The sequential development of the cortical arteries resulting in a layered arterial supply may depend upon the cytoarchitectonic lamination of the cerebral cortex. Variations in the cortical angioarchitecture seem to be related to an area- and lamina-specific growth of the cerebral cortex (Lierse 1963; van den Bergh 1965). A *tangential* growth of the cortex increases the mean distance between the vertically penetrating vascular trunks. An ingrowth of new intracortical branches of leptomeningeal arteries occurs, which maintains an adequate supply to the newly developing layers. After this process has been terminated during the second week after birth, the following tangential growth of the cortex leads to a decrease in the packing density of the vertically oriented trunks.

A hexagonal-like spatial distribution of the vertically oriented intracortical arteries and veins is observed in sections cut parallel to the pial surface. In adult rats the intracortical vascular trunks can be divided into three size classes according to their caliber, depth of penetration, and terminal territory of the vessel (Wolff 1976a). The earliest vessels which run through all cortical layers and terminate in the presumptive subcortical structures represent the largest units of vascular supply (vascular module). The later-developed cortical trunks are arranged concentrically around the subcortical trunks showing decreasing distances from each other according to their time of origin and their caliber respectively. A comparable pattern of distribution was described by Christaller (cited by Schöller 1972) concerning the spatial arrangement of central places in geographic systems, which is characterized by an optimal relation between the number of central places of supply and their distances from each other (Schöller 1972). In the cerebral cortex the topological relations of the different vascular trunks are influenced by the specific growth and differentiation of the neuronal tissue. The angioarchitectonic of the cerebral cortex is built up by vascular modules of different size which are successively arranged in layers during development.

4.3 Second Stage of Intracortical Vascularization

The development of new intracortical vascular branches originating from distal parts of the leptomeningeal net ceases during the second postnatal week. The intracortical terminal vessels, however, continue to develop new branches until the end of the third postnatal week (Bär and Wolff 1973). In the neonatal period a basic cortical angioarchitecture is established by hexagonal-like packed vascular trunks which are interconnected by terminal vessels forming arcades and wide meshes. These immature vessels can obviously develop into different parts of the terminal vascular bed (metarterioles, arteriovenous thoroughfare channels, net-capillaries, postcapillary venules). During the second stage of intracortical vascularization an extensive proliferation of net-capillaries takes place leading to lamina- and area-specific differences in capillary density (Craigie 1920, 1925; Bär 1978). The net-capillaries represent a fraction of about 90% of all vascular profiles counted in the adult cerebral cortex.

During the first postnatal week the initial proliferation of the definitive capillaries leads to an increase in branching density within the terminal vascular net, which is followed by a marked increase in L_v a few days later. The postnatal increase in L_v coincides in time with an increasing fraction of seamless capillary tubes (Bär and Wolff 1977). The formation of such monoendothelial tubes, which do not show any interendothelial contact zones in their cross sections, apparently occurs at the fusion point of capillary sprouts with preexisting vessels (Wolff et al. 1975). That assumption is supported by the parallelism between the increasing number of seamless endothelial cells and the increase in L_v, which makes the fusion of sprouts with preexisting vessels more probable.

Besides the collective stereological parameters, the cellular composition of the microvascular bed changes too. During the first days after birth a marked increase in the volume of the neocortex takes place. The numeric density and the mean length of endothelial cells remain rather constant during that period. The branching density of the capillary net, however, increases rapidly during the following days. That means that new capillary branches are formed by a large fraction (about 60%) of all endothelial mitoses. The endothelial cells continue to divide rapidly a few days after the onset of saturation of the cortical volume growth. The resulting increase in the specific capillary length is accompanied by an increase in both the numerical density and the mean length of endothelial cells (Table 6).

The postnatal vascularization coincides in time with a marked increase in brain blood flow. With increasing age the blood flow of various cerebral structures rises from relatively low levels to peak levels and then gradually declines as maturity is achieved (Kennedy et al. 1972). The changes of cortical blood flow seem to be correlated with an altered relation between capillary length and branching density of the microvascular net (Bär and Wolff 1973), which may cause changes in the resistance of flow and in the perfusion pattern.

4.4 Third Stage of Intracortical Capillarization

The late stage of capillary growth is characterized by a decreasing mitotic activity of the endothelial cells. During this period the final one-third of the collective length of capillaries in the adult cortex and the final 15%—20% of the capillary density in layer IV is formed. The numerical density of endothelial cells remains nearly constant. The further increase in L_v is mainly caused by an elongation of the existing endothelial cells. During the third stage of intracortical capillarization passive changes in L_v and in branching density take place, accompanying the further growth of volume of different cortical laminae. For instance, in the basal cortical layers the concentration of the existing capillaries is attenuated in an increasing tissue volume caused partly by myelination of axons. A hypervascularization preceding the myelination has been observed in other parts of the nervous tissue too (Kennedy et al. 1970). The observed morphological alterations of the capillary net may correspond to the changing local blood flow during comparable periods of the postnatal development (Kennedy et al. 1970).

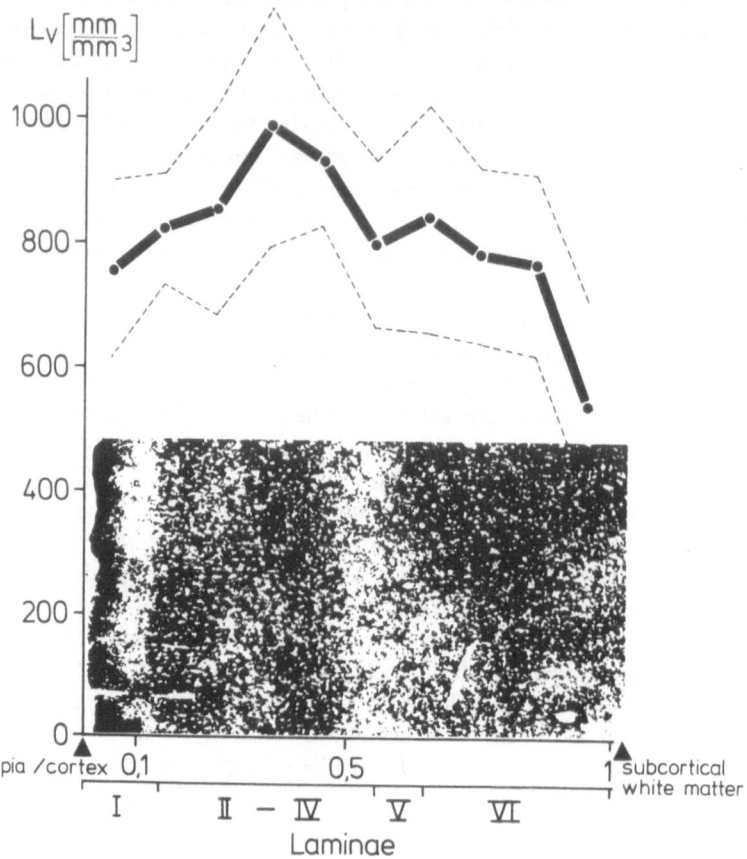

Fig. 33. Specifc capillary length (L$_V$) (at six months) and density of mitochondria in different layers of the rat's visual cortex. Capillaries and mitochondria are distributed in corresponding bands. Note the presence of the same peaks and valleys. (Silver impregnation of mitochondria; the micrograph was kindly provided by Prof. J.R. Wolff). The *lower scale* presents the cytoarchitectonic laminae

   The morphological and functional maturation of different brain regions is accompanied by an increase in L$_v$ which takes place in a temporal pattern characteristic for each special region. The proliferation of capillaries is associated in time with an increasing number of synaptic contacts (Aghajanian and Bloom 1967; Wolff 1976b, 1978) and of mitochondria (Pysh 1970). A close correlation between L$_v$ and the numerical density of synaptic contacts does not exist, because the distribution of synapses depends upon the intrinsic neuronal organization and the pure number of synapses ist not sufficient to evaluate their functional activity. The capillary density, however, may vary according to topological factors such as the spatial relation to penetrating arterioles and venules. A closer correlation may exist between L$_v$ and the density of mitochondria as suggested by Scharrer (1945, 1962) and by the present results (Fig. 33). However, as yet there are no detailed quantitative studies concerning that question.

4.5 Vascular Sprouting

An effective growth of the capillary net depends upon an induction of endothelial mitosis and subsequent sprout formation. Chemical growth factors stimulating mitotic division and migration of endothelial cells are mediated during ontogeny, wound healing, and tumor expansion. Detailed knowledge about the chemical composition of such vasoactive factors is lacking. Capillary proliferation can be induced by angiogenic factors produced by some tumors (Folkman and Cotran 1976, Cavallo et al. 1972). One can only speculate that similar factors may act during ontogeny (Diemer 1968, Finkelstein et al. 1977) and under hypoxic and ischemic conditions. A mathematical model by which the initial vascular pattern of a melanoma transplant can be simulated is reported by Deakin (1976). The author assumes that the tumor produces an angiogenic factor (TAF) by which sprouts can be elicited from the host vessels. These sprouts then invade the tumor along the gradient of TAF. The ability to develop new sprouts depends upon the stage of maturity of the vessel wall and the adjacent tissue. Growing vascular networks consist of a homogenous population of *undifferentiated* endothelial cells (Bremer 1960; Ausprunk et al. 1974) which form capillaries without a continuous basal lamina (Aloisi and Schiaffino 1971; Bär and Wolff 1972). A well-developed basal lamina may prevent the migration of endothelial cells and may diminish the action of a growth factor. This would agree with the observation that the formation of sprouts is restricted to the undifferentiated parts of the vascular system.

The development of the organ-specific contact relations between the capillary wall and the surrounding tissue has to be included among the factors that control the vascular growth. During the postnatal development of the cortex the decreasing mitotic activity of endothelial cells is associated in time with the maturation of perivascular structures (Bär and Wolff 1976). The termination of the vascular sprouting is morphologically characterized by a continuous basal lamina (thickness: 30–40 nm) and a perivascular astroglial sheath which completely separates the vascular wall from the neuronal tissue. A potential induction of vascular growth depends upon the degree of structural and functional maturation of the neuronal tissue, which is accompanied by the formation of lamellar parts of astroglial cells (Wolff and Bär 1976) and synaptogenesis (Wolff 1976b; Vrensen et al. 1977). In young brains the oxidative utilization of glucose is limited. The increasing functional activity of the cerebral cortex is accompanied by changes in the metabolism (Land et al. 1977). During that period an accumulation of some metabolites of anaerobic glycolysis caused by a low capacity of the oxidative metabolism may be responsible for the initiation of the rapid proliferation of capillaries. This would agree with the observed parallelism between the postnatal changes of branching density and the regional O_2-consumption (Bär and Wolff 1973; Warshaw and Terry 1976). Vascular adaptation to the increasing demand for O_2 by mitotic proliferation takes place during a limited period of normal ontogeny. Later vascular growth is characterized by changes in size and shape of a relatively constant population of endothelial cells.

4.6 Effects of Hypoxia upon Postnatal Capillarization

The vulnerability of the CNS against O_2-deprivation depends upon the ontogenic phase during which it is applied. Newborn rats can survive a complete anoxia longer than adults (Fazekas et al. 1941). The low O_2-consumption corresponds to the limited ability of the young brain to utilize glucose for its energy requirements. Nevertheless, neonatal rats develop a retardation of growth and reductions of their brain weights and DNA, protein, and RNA contents after being exposed to 12% oxygen from 1 to 7 days (Cheek at al. 1969). A similar aerogenic hypoxia has no marked influence on the thickness of the occipital cortex when applied from days 4 to 14 after birth. The myelination of the basal cortical layers and the size of the perikarya of lamina V pyramidal cells are retarded after hypoxia. It is suggested that the resulting disturbances of the neuronal tissue cannot be prevented by vascular adaptation. Nevertheless, the volume fraction of terminal vessels is increased from 1.1 vol.% in 14-day-old controls to 1.7 vol.% in animals after hypoxia during days 4 to 14.

After hypoxia the percentage of ^{3}H-thymidine-labeled cells is increased as compared with controls of the same age. The observed dilation of the vascular bed is not accompanied by a flattening of the vascular wall. This favors the assumption that not only an increase in the turnover rate but also enlargement of the existing and intercalation of additional endothelial cells have occurred. The changes of the morphological parameters of the brain capillaries disappeared after survival of the hypoxic animals except for the volume fraction of the capillaries, which was diminished to 1.4 vol.% as compared with 1.7 vol.% in normal animals of the same age. A reduced volume fraction of capillaries is combined with a reduction of regional blood volume which is correlated to a decrease in regional blood flow as suggested by Risberg et al. (1969). One possible cause of the observed changes in cortical blood supply may be related to a disturbed pattern of synaptogenesis induced by hypoxia which results in an altered relation of two different populations of synapses (Bär 1977) separated by morphological means according to the criteria given by Gray (1971). Immediately after hypoxia from days 4 to 14, type I synaptic contacts in the visual cortex are reduced in number by about 25% (Bär 1977). The numerical density of type II synapses was not altered on day 14. After survival of the hypoxic animals until day 55 after birth the density of type II synapses remains at a low level, whereas the density of type I synapses no longer shows any significant differences from controls of the same age (Bär 1977). After hypoxia during the early postnatal period, a long-lasting disturbance of neuronal circuitry results; this may be responsible for the detected change in the size distribution of capillary diameters.

Thus, hypoxia during the period of maximal cortical growth does not result in gross morphological changes. This is completely different in the following developmental period, which is characterized by activation of the oxygen-dependent energy metabilism and increase in oxygen consumption (Land 1977). During this period, described as critical by Flexner (1952), cytologic and functional changes take place in the cerebral cortex. The process of cortical maturation is accompanied by a maximal proliferation of capillaries, whereas the increase in neocortical volume reaches a plateau during the same period. After hypoxia from days 14 to 54, a 14% reduction in the thickness of the visual cortex is observed, which corresponds to a decrease in volume of about 42% provided that the reduction takes place in three dimensions equally. After hypoxia the thickness of the visual cortex had fallen below its level on day 14 when the experiment was started. The observed increase in L_v is probably not caused by active growth of capillaries, but by passive concentration of the *existing* capillary net in a tissue volume reduced by atrophy.

The present results do not concern the time of onset and the progression of the hypoxic changes. However, if structural integrity of perfused brain capillaries is assumed and no additional capillarization is induced by hypoxia, the cortical atrophy caused by the hypoxic stress starts on approximately day 20 after birth. During normal development a corresponding value of L_v is reached between days 20 and 30 after birth. Provided that a 42% reduction of the cortical volume and no further capillary growth have occurred until day 55, the resulting value of L_v corresponds to the capillary density found in the visual cortex of hypoxic animals at day 55.

Thus, the *course* of maximal brain growth is not markedly influenced by oxygen deprivation. The special vulnerability of the cerebral cortex may be explained by a disturbance of the ordered sequential pattern of biochemical and functional maturation. During exposure to a N_2-O_2 environment the rate of synthesis and degradation of catecholamines and 5-hydroxytryptophan is decreased (Hedner 1978). The activity of tyrosine hydroxylase, tryptophan hydroxylase, and monoamine oxidase have generally been considered to reflect the functional maturity of the brain monoamine pathway. The vulnerable period (between days 14 and 30) of brain development is characterized by a maximal rate of myelination (Davison and Dobbing 1966; Haug et al. 1976). At the same time several neurophysiologic (evoked potentials, latencies, fatigability) and biochemical parameters (transendothelial transport, energy metabolism) reach the adult pattern (Gramsbergen 1976; Mareš and Vítová 1973; Verley and Axelrad 1975; Cremer et al. 1976; Warshaw and Terry 1976, Land et al. 1977).

The developmental pattern of some enzymes of the energy metabolism may be influenced by nutritional factors too. During periods of high fat intake in suckling rats the brain has an increased capacity for long chain fatty acid oxidation (Warshaw and Terry 1976). This ontogenic period is characterized by the high activity of ketone body-utilizing enzymes (Dahlquist et al. 1972). Ketone bodies are used as substrates for the synthesis of lipids during myelination and are metabolized together with fatty acids as alternative substrates to meet the energy requirements of the developing brain during the period of limited capacity of oxidative utilization of glucose. During weaning, which normally begins at 20 days of age, pyruvate dehydrogenase reaches its adult activity (Land et al. 1977), whereas the activity of enzymes that participate in the utilization of ketone bodies (e.g., D-β-hydroxybutyrate dehydrogenase) shows a decreasing trend (Warshaw and Terry 1976). The decreasing utilization of ketone bodies together with the decreased food intake caused by the hypoxic stress favor the development of ketoacidosis in young rats. The activity of pyruvate dehydrogenase, however, is inhibited by the increased levels of ketone bodies. Thus, the oxidative utilization of glucose via the tricarboxylic acid cycle is retarded during hypoxia. It is suggested that an efficient and active aerobic glycolyses is necessary for full neural expression (Land et al. 1977), including functional and biochemical maturation of neurons. Therefore, the atrophy of the cerebral cortex observed after hypoxia may be caused by disturbances of the normal ontogenic pattern of enzymatic adaptation to glucose oxidation.

Hypoxia can be compensated without atrophy when applied after the critical period in the development of the energy metabolism. Hypoxia from days 30 to 70 results in a 5% loss of brain weight and a 2% reduction in the cortical thickness as compared with controls of the same age. An effective vascular adaptation occurs by intercalation of additional endothelial cells in the existing capillaries (resulting in an increase in L_v of about 20%) during this stage of development. The branching density of the capillary net, however, remains unchanged under the applied hypoxic conditions. The volume fraction of the terminal vessels represents 4.5% (s = ± 1.9) of the cortex volume in hypoxic and 2.2% (s = ± 0.4) in controls of the same age. Structural and functional alterations of the cerebral cortex, which may accompany the observed increase in capillary blood volume, remain to be evaluated.

After the first postnatal month the rat's increasing tolerance to hypoxia is associated with the establishment of the adult pattern of mitochondrial enzymes (Land et al. 1977) and of the adult electroencephalogram (Gramsbergen 1976). During the same period the maturation of several transport processes between blood and brain tissue is achieved (Sessa and Perez 1975; Cremer et al. 1976). The brain covers its energy requirements almost exclusively by glucose oxidation. The blood flow (Kennedy et al. 1972) and the oxygen consumption of the brain (Tyler and van Harreveld 1942) gradually decrease from peak values between 20 and 30 days after birth to adult values.

A lowered partial pressure of oxygen in the inspired air results in different disturbances of brain development depending upon the stage of maturation at the beginning of the experiment (Büchner 1975). Hypoxia affects the ordered temporal and spatial sequence of functional and biochemical maturation of the cortical tissue when applied between days 14 and 30 after birth. Thus, the metabolic conditions leading to a normal vascularization are changed. The reversibility of the observed changes in cortical development may partly depend upon the limited growth of the vascular system because of the termination of the sprouting process during the early postnatal period. The adult intracerebral capillary system shows only small plastic deformations of the existing endothelial cells during adaptation to different hemodynamic conditions (Bär et al. 1975; Mayrovitz and Wiedeman 1975; Petito et al. 1977) including hypertension (Nag et al. 1977, Bohlen et al. 1977). The increased mitotic activity of the capillary endothelial cells in the brain of spontaneously hypertensive rats (Hazama et al. 1977) may be an expression of repair processes of the injured endothelial cells rather than an indication of an increase in the number of endothelial cells.

4.7 Intracortical Capillaries During Aging

The aging process at the cellular level is characterized by disturbances in DNA and RNA synthesis (Ono and Cutler 1978) and by a progressive lengthening of cell generation time (Christofalo and Sharf 1973; Sheldrake 1974; Mareš and Lodin 1974; Andrews et al. 1976; Chetsanga et al. 1977). Due to the suggested age-dependent accumulation of defects in the genome (Ono and Cutler 1978) some new cells, which differ from their original configuration, are generated (Samaras 1974; Chetsanga et al. 1977). In the mouse forebrain a decreasing tendency of DNA synthesis can be observed in glia and endothelial cells with advancing age (Mareš and Lodin 974: Dalton et al. 1968). Cell degeneration and the resulting cell loss mainly involve the neurons (Brody 1976, Bugiani et al. 1978). The glia cells, however, can be renewed and represent a population that reacts in a different way in different regions. In the aging brain the number of glia cells is increased in some brain regions (Vaughan and Peters 1974; Brizee et al. 1968; Vernadakis 1975). The turnover rate of the endothelial cells, which is commonly very low in the CNS compared to other organs (Engerman et al. 1967; Korr et al. 1975), may be influenced by the aging process. At present, one can only speculate about the possible consequences associated with a disturbed equilibrium of the population of endothelial cells.

The aging process is accompanied by *regressive* changes of the whole vascular system. A quantitative loss of cells in the media layers of arteries and veins, which leads to a replacement of the degenerating smooth muscle cells by collagenous connective tissue, has been described (Nanda and Getty 1971, Iwanowski 1974; Fang 1976). Atrophic age changes combined with hypertrophia may occur in the walls of small vessels (Fang 1976). The mean length of intracortical capillary endothelial cells (L_E) increases during aging as the present results show. There is a further indication that the cellular composition of the aging microvascular bed has changed. The mean specific length of brain capillaries (L_v) increases during aging. This can be interpreted as a concentration of the *existing* capillary net in a smaller tissue volume caused by brain atrophy. Thus, a proliferation of new capillaries is not a prerequisite for the observed increase in L_v and the degenerative changes may predominate the proliferative processes in the aging microvascular net. This assumption agrees with the observation that the increased L_v is not accompanied by an adequate increase in endothelial cell density. There are fewer endothelial cells per unit length of capillaries in 30-month-old animals than in 23-month-olds. The increasing number of intercapillary bridges without an open vascular lumen (Guseo and Gallyas 1974) can be considered a further morphological sign of capillary regression (Ausprunk et al. 1976). Indeed, a decrease of the mean diameter of brain microvessels occurs, which agrees with the observation of Vaughan and Calvin (1977) who found a decreasing blood volume in the aging brain.

The senile changes of the intracortical capillary net are inhomogeneously distributed over the cortical laminae. The reduction of the mean diameter of rat brain capillaries was at first detected in layers 3, 4, 6 and 7 which correspond to the cytoarchitectonic laminae III and V. The observed vascular alterations which impair the regional microcirculation may be associated with neuronal age changes. For example, in clinical cases of presenile dementia a reduction of the regional cerebral blood flow was suggested to be related to a local loss of neurons (Johannesson et al. 1977).

The question whether there is a causal relation between neuronal and vascular changes is very difficult to answer because of methodological problems. The results of cell counting in histologic sections, for example, are influenced by the species examined, by different anamnestic data (Shefer 1973, Colon 1973), by differences in the brain region, and by the methods chosen for the morphometric measurements (Tomasch 1972; Haug 1975; Brody 1976; Bugiani et al. 1978). The results of Colon (1972) favor the assumption of a neuronal cell loss that is homogeneously distributed over the whole depth of the cerebral cortex, whereas Shefer (1973) found a nerve cell loss restricted to lamina III. However, the degeneration of the structural elements of the neurons such as dendritic and/or axonal arborizations including synapses need not coincide in time and space with a possible loss of neurons. Thus, several authors (Scheibel et al. 1975; Feldman and Dowd 1976; Hinds and McNelly 1977; Geinisman et al. 1977) found a pattern of involution specific for each cortical region and lamina. In the cerebral cortex the horizontally oriented dendritic systems of the pyramidal cells and the corresponding spine population seem to be preferentially affected by the aging process. This is in accordance with the observed vascular changes which begin in those cortical layers showing a marked involution of the neuropil (Scheibel et al.

1975; Feldman and Dowd 1976). The aging process includes general organ-specific and individual factors and is based upon cellular alterations. We do not know the localization of the initial steps for the senile changes: whether aging begins at the level of peripheral receptors or whether it is induced by intrinsic changes at the synaptic terminal (Bowen et al. 1974; Feldman and Dowd 1976; White et al. 1977). The endothelial cells of the brain play a more important role than hitherto suspected during a developmental period characterized by a generally decreasing mitotic activity of proliferating cell systems (Silini and Andreozzi 1974).

The position of the endothelial cell line emphasizes its important role for the transport process between blood and brain tissue which is necessary to maintain the microenvironment of the neurons. Ultrastructural changes of the capillary wall, e.g., thickening of the basal lamina (Bär and Strauch 1979) suggest an impairment of the permeability and the elastic properties of the capillaries (Knobloch 1956; Krebs and David 1962). Apart from the morphological changes there are age-dependent changes of the enzyme activity in the endothelial cells which may play a role in the maintenance of the blood-brain barrier (Panula and Rechardt 1978). Local disturbances of the blood-brain barrier may induce a proliferation of glia cells as described in certain brain regions (Brizzee et al. 1968; Vernadakis 1975; Vaughan and Calvin 1977); The biochemical error may at first develop in the glia and may subsequently influence the neurons as suggested by Vernadakis (1975).

In further studies directed to the developmental sequence of the involution process, the endothelial cells of brain microvessels and a possible causal relation between vascular and neuronal alterations should be considered.

5 Summary

The vascular system of the occipital cortex of 137 albino rats was studied using morphometric methods. The developmental changes of the mean numerical density (N_E), the mean length (L_E), and the mean arithmetic thickness of endothelial cells were determined to examine the cellular composition of the microvascular system. In addition, the interendothelial contact relation and the pericapillary astroglial sheath were quantitatively evaluated. In order to describe the process of vascularization, the following morphometric parameters were determined by automatic image analysis of capillary profiles in semithin sections: specific capillary length (L_V), volume fraction of capillaries (V_V), and specific internal surface area of capillaries (S_V), as well as the size distribution of the internal vascular diameters.

Vascular growth includes two different processes: increase in endothelial cell number by mitotic division and elongation of the existing endothelial cells. Mitosis of endothelial cells results in the formation of new vascular sprouts or in intercalation of additional cells within the vascular wall. The mitotic activity of endothelial cells decreases with advancing age. Therefore capillary growth is restricted to an elongation of single endothelial cells as maturity is reached.

During ontogeny the intracortical vascular tree is developed in three successive stages which differ in pattern and localization of the vascular growth process.

1. First Stage. All vessels that supply the neocortex sprout from the perineural, leptomeningeal capillary net. While the growth in surface and thickness of the cortex occurs, radially penetrating vascular sprouts, which later become the radially oriented trunks, grow into the neural tissue. This process is terminated in a temporal and spacial pattern which corresponds to the maturation of the leptomeningeal vascular walls. The surface area of the neocortex continues increasing after this, causing a decrease in the packing density of the vertically oriented vascular trunks during the late postnatal growth of the cortex. The existing vascular trunks are elongated when the cortical thickness increases. Associated with the inside-out layering of pyramidal neurons and with the sequence of maturation of horizontal cortical laminae, the newly ingrowing vessels only reach the more superficial layers. Thus, the different length of the radially penetrating vessels and, accordingly, the different sizes of their branching areas are determined by the special development of the cortical tissue.

In the neocortex the radially penetrating vessels are distributed almost hexagonally, according to the "system of central places", the hierarchy in the system being determined by the vascular length, which permits conclusions as to the date of origin of the vessels. The oldest vessels are the longest (rami medullares) supplying the largest areas. The younger radially penetrating vessels are shorter (rami corticales) and more numerous with smaller branching areas.

2. Second Stage. Capillary sprouts that originate from the terminal intracortical vessels build up the definitive capillary net. The capillaries represent about 90% of the collective intracortical vascular length. They are formed by a single row of elongated hexagonal endothelial cells enveloping the lumen. The hexagonal shape and the arrangement in monocellular sequence agree with the occurrence of twice as many capillary cross sections with one endothelial cell junction (ECJ) as with two ECJs. The fraction of seamless endothelial tubes amounts to about 30% of all cortical capillaries. The frequency of seamless capillaries is suggested to represent a characteristic feature of late-developing capillary networks.

The formation of new intracortical capillary sprouts is finished at the same time as the morphological maturation of the intracortical capillary walls.

3. Third Stage. The late stage of capillary growth is characterized by a decreasing mitotic activity of the endothelial cells. The further increase in L_V is mainly due to passive elongation of the existing endothelial cells and produces the final one-third of the collective length in the adult cortex. The numerical density of endothelial cells is kept nearly constant. The regional- and lamina-specific differences in L_V which are developed during the second and third stage of capillarization seem to be dependent upon the functional maturation of the cortical tissue.

Oxygen Deficiency

Vascular adaptation is influenced by the extent of O_2-deficiency and by the developmental stage of the cortex. Dilation of vessels is generally evoked, accompanied by an increase in size of endothelial cells. If O_2-deprivation starts at an early stage of postnatal development the maturation of the neocortex is retarded. The thickness of the cortex, however, is not significantly changed. An increased ^{3}H-thymidine labelling index does not result in a higher branching density of the capillary net.

The cortical tissue is especially sensitive to hypoxia when its own O_2 demand is particularly high. This coincides in time with the maximal rate of myelination and with changes in the pattern of oxidative enzyme systems. In the neocortex of the rat these processes take place at the end of the third postnatal week. Hypoxia at this critcal stage causes a decrease in cortical thickness associated with a passive increase in L_V. Thus, the severe developmental disorder cannot be averted by vascular adaptation.

However, an equivalent O_2-deprivation is compensated without cortical atrophy when applied from days 30 to 70. Vascular dilation and a distinct increase in L_V are observed in the occipital cortex.

In adult rats O_2-deficiency alone is not an adequate stimulus to cause capillary proliferation. In the intracortical vascular bed of adults, chronic hypoxia causes only vascular dilation, which is apparently the consequence of the altered balance between intravascular and tissue pressure. Nevertheless, L_v remains nearly constant. Therefore it does not seem very likely that there are *reserve* capillaries in the cortex which are closed under normal conditions. Obviously, the development of new capillary branches in the adult brain is only possible if there is a disturbance of the structural continuity of the tissue.

Aging

The earliest age changes of the microvascular bed concerning the size distribution of vessel calibers are detected in the cytoarchitectonic laminae III and V. This laminar pattern corresponds to the distribution of initial degenerative changes of neuronal elements.

With advancing age the thickness of the neocortex decreases; especially laminae II—IV are reduced. This coincides with an adequate increase in L_V. L_E increases during senescence. The increase in L_V is not accompanied by a corresponding increase in the number of endothelial cells per unit tissue volume. There are fewer endothelial cells per unit length of capillaries in 30-month-old animals than in 23-month-olds. During aging an endothelial cell loss may occur; it can be partly compensated by elongation of the existing endothelial cells.

References

Aghajanian GK, Bloom FE (1967): The formation of synaptic junctions in developing rat brain: a quantitative electron microscopic study. Brain Res 6: 716–727

Allen N, Clendenon NR, Abe H, Swenberg JA, Koestner A, Wechsler W, Shuttleworth EC jr (1977) Acid hydrolase and cytochrome oxidase activities in nitrosourea induced tumors of the nervous system. Acta neuropathol (Berl) 39: 13–23

Aloisi M, Schiaffino S (1971) Growth of elementary blood vessels in diffusion chambers. II. Electron microscopy of capillary morphogenesis. Virchows Archiv (Cell Pathol) 8: 328–341

Andrews AD, Barrett S, Robbins JH (1976) Relation of DNA repair process to pathological ageing of the nervous system in xeroderma pigmentosum. Lancet 7973: 1318–1321

Ashton N, Tripathi B, Knight G (1972) Effect of oxygen on the developing retinal vessels of the rabbit. Exp Eye Res 14: 214–220

Ausprunk DH, Knighton DR, Folkman J (1974) Differentiation of vascular endothelium in the chick chorioallantois: a structural and autoradiographic study. Dev Biol 38: 237–248

Ausprunk DH, Falterman K, Folkman J (1976) The sequence of events in regression of capillaries. J Cell Biol 70: 209a

Bär Th (1977) Wirkung chronischer Hypoxie auf die postnatale Synaptogenese im Occipitalcortex der Ratte. Verh Anat Ges 71: 915–924

Bär Th (1978) Morphometric evaluation of capillaries in different laminae of rat cerebral cortex by automatic image analysis: changes during development and aging. In: Cervos-Navarro J, Betz E, Ebhardt G, Ferszt R, Wüllenweber R (eds) Pathology of cerebrospinal microcirculation. Raven Press, New York (Adv Neurol Vol 20, pp 1–9)

Bär Th, Strauch L (1979) Messungen der Kapillarwanddicke im Cerebralcortex alternder Ratten. Verh Anat Ges 73: 1069–1073

Bär Th, Wolff JR (1972) The formation of capillary basement membranes during internal vascularization of the rat's cerebral cortex. Z Zellforsch mikrosk Anat 133: 231–248

Bär Th, Wolff JR (1973) Quantitative Beziehungen zwischen der Verzweigungsdichte und Länge von Capillaren im Neocortex der Ratte während der postnatalen Entwicklung. Z Anat Entwicklungsgesch 141: 207–221

Bär Th, Wolff JR (1976) Development and adult variations of the wall of brain capillaries in the neocortex of rat and cat. In: Cervos-Navarro J, Matakas F (eds) The cerebral vessel wall. Raven Press, New York, pp 1–6

Bär Th, Wolff JR (1977) Morphometry of interendothelial and glio-vascular contacts of rat brain capillaries during postnatal development. Bibl Anat 15 I: 514–517

Bär Th, Wolff JR, Hunziker O (1975) Effects of different hemodynamic conditions on brain capillaries: alveolar hypoxia, hypovolemic hypotension, and ouabain edema. In: Penzholz H, Brock M, Hamer J, Klinger M, Spoerri O (eds) Brain hypoxia-pain. Springer, Berlin Heidelberg New York Adv Neurosurg Vol 3, pp 10–19

Baez S (1977) Microcirculation. Ann Rev Physiol 39: 391–415

Balashova ET (1956) Einige Ergebnisse über die Nachbarschaftsbeziehungen zwischen Nervenzellen und Kapillaren im Trigeminuskern während der postnatalen Entwicklung. Biull Eksp Biol Med (Moskau) 42: 71–74

Bergh van den R (1965) Einige Besonderheiten der intracerebralen Gefäßanordnung. Zentralbl Neurochir 25: 180–197

Bergh van den R, Eecken van der H (1968) Anatomy and embryology of cerebral circulation. Prog Brain Res 30: 1–25

Bohlen HG, Gore RW, Hutchins PhM (1977) Comparsion of microvascular pressures in normal and spontaneously hypertensive rats. Microvasc Res 13: 125–130

Bowen DM, White P, Flack RHA, Smith C, Davison AN (1974) Brain-decarboxylase activities as indices of pathological change in senile dementia. Lancet 7869: 1247–1248

Bremer H (1960) Untersuchungen über den formalen Zusammenhang zwischen einer Struktur und deren Entwicklung am Gefäßnetz der Hühnerkeimscheibe. Arch Entwicklungsmech Org (Wilhelm Roux) 152: 585–592

Brizzee KR, Sherwood N, Timiras PS (1968) A comparison of cell populations at various depth levels in cerebral cortex of young adult and aged long-evans rats. J Gerontol 23: 289–297

Brody H (1976) An examination of cerebral cortex and brainstem aging. In: Terry RD, Gershon S (eds) Neurobiology of aging. Raven Press, New York (Aging Vol 3, pp 177–181)

Büchner F (1975) Die Entwicklung des Embryo bei normalem und gestörtem Stoffwechsel. In: Büchner F (ed), Hypoxie. Springer, Berlin, Heidelberg, New York, pp 449–469

Burgiani O, Salvarani S, Perdelli F, Mancardi GJ, Leonardi A (1978) Nerve cell loss with aging in the putamen. Eur Neurol 17: 286–291

Bull JWD (1975) The nervous system's blood vessels from Galen to Röntgen and after. Proc Soc Med 68: 695–702

Burian WG (1970) Der Einfluß chronischen Sauerstoffmangels auf die Kapillardichte in der Groß-hirnrinde erwachsener Ratten. Inaugural Diss, Faculty of Medicine, Munich

Caley DW, Maxwell DS (1970) Development of the blood vessels and extracellular spaces during postnatal maturation of rat cerebral cortex. J Comp Neurol 138: 31–48

Camosso ME, Roncalli L, Ambrosi G (1976) Vascular patterns in the chick embryo spinal cord in normal and experimentally modified development. Acta Anat (Basel) 95: 349–367

Campbell ACP (1939) Variation in vascularity and oxidase content in different regions of the brain of the cat. Arch Neurol Psychiatr 41: 223–242

Cavallo T, Sade R, Folkman J, Cotran RS (1972) Tumor angiogenesis. Rapid induction of endo-thelial mitoses demonstrated by autoradiography. J Cell Biol 54: 408–420

Cheek DB, Graystone JE, Rowe RD (1969) Hypoxia and malnutrition in newborn rats; effects on RNA, DNA, and protein in tissues. Am J Physiol 217: 642–645

Chetsanga CJ, Tuttle M, Jacoboni A, Johnson C (1977) Ageassociated structural alterations in senescent mouse brain. Biochem Biophys Acta 474: 180–187

Cobb S, Talbott JH (1927) Studies on cerebral circulation. II. Quantitative study of cerebral capillaries. Trans Assoc Am Physicians 42: 255–272

Cohnheim J (1872) Untersuchungen über die embolischen Prozesse. Berlin

Colon EJ (1972) The elderly brain. A quantitative analysis in the cerebral cortex of two cases. Psychiatr Neurol Neurochir 75: 261–270

Colon EJ (1973) The cerebral cortex in presenil dementia. A quantitative analysis. Acta Neuro-pathol (Berl) 23: 281–290

Craigie EH (1920) On the relative vascularity of various parts of the central nervous system of the albino rat. J Comp Neurol 31: 429–464

Craigie EH (1925) Postnatal changes in vascularity in the cerebral cortex of the male albino rat. J Comp Neurol 39: 301–324

Craigie EH (1945) The architecture of the cerebral capillary bed. Biol Rev 20: 133–146

Cremer JE, Braun LD, Oldendorf WH (1976) Changes during development in transport processes of the blood-brain barrier. Biochim Biophys Acta 448: 633–637

Cristofalo VJ, Sharf BB (1978) Cellular senescence and DNA synthesis. Thymidine incorporation as a measure of population age in human diploid cells. Exp Cell Res 76: 419–428

Dahlquist G, Persson U, Persson B (1972) The activity of D-ß-hydroxybutyrate dehydrogenase in fetal, infant and adult rat brain and the influence of starvation. Biol Neonate 20: 40–50

Dalton MM, Hommes OR, Leblond CP (1968) Correlation of glial proliferation with age in the mouse brain. J Comp Neurol 134: 397–399

Davison AN, Dobbing J (1966) Myelination as a vulnerable period in brain development. Br Med Bull 22: 40–44

Deakin AS (1976) Model for inital vascular patterns in melanoma transplants. Growth 40: 191–201

Diemer K (1963) Eine verbesserte Modellvorstellung zur Sauerstoffversorgung des Gehirns. Natur-wissenschaften 50: 617–618

Diemer K (1968) Capillarisation and oxygen supply of the brain. In: Lübbers DW, Luft UC, Thews G, Witzleb E (eds) Oxygen transport in blood and tissue. Thieme, Stuttgart, pp 118–123

Duret H (1874) Recherches anatomiques sur la circulation de l'encéphale. Arch Phys Norm Path 2: 60–91; 316–354; 664–693; 919–957

Eayrs JT (1954) The vascularity of the cerebral cortex in normal and cretinous rats. J Anat 88: 164–173

Eins S, Bär Th (1978) Orientation distribution of blood vessels in central nervous tissue. In:

Quantitative analysis of microstructures. Sonderbände der Praktischen Metallographie Bd 8. Riederer, Stuttgart, pp 381–388

Eins S, Wilhelms E (1976) Assessment of preparative volume changes in central nervous tissue using automatic image analysis. Microscope 24: 29–37

Engerman RL, Pfaffenbach D, Davis MD (1967) Cell turnover of capillaries. Lab Invest 17: 738–743

Fang HCH (1976) Observations on aging characteristics of cerebral blood vessels, macroscopic and microscopic features. In: Terry RD, Gershon S (eds) Neurobiology of aging. Raven Press, New York (Aging Vol 3, pp 155–166)

Fazekas JF, Alexander FAD, Himwich HA (1941) Tolerance of the newborn to anoxia. Am J Physiol 134: 281–287

Feeney J jr, Watterson RL (1946) The development of the vascular pattern within the walls of the central nervous system of the chick embryo. J Morphol 78: 231–202

Feldman ML, Dowd C (1976) Loss of dendritic spines in aging cerebral cortex. Anat Embryol (Berl) 148: 279–302

Finkelstein D, Brem St, Patz A, Folkman J, Miller St, Ho-chen Ch (1977) Experimental retinal neovascularization induced by intravitreal tumors. Am J Ophthalmol 83: 660–664

Flexner LB (1952) The development of the cerebral cortex. A cytological, functional and biochemical approach. Harvey Lect 47: 156–179

Folkman J, Cotran R (1976) Relation of vascular proliferation to tumor growth. Int Rev Exp Pathol 16: 207–248

Friede RL (1966) Capillarization. In: Friede RL (ed) Topographic brain chemistry. Academic Press, New York, pp 1–15

Geinisman Y, Bondareff W, Dodge JT (1977) Partial deafferentation of neurons in the dentate gyrus of the senescent rat. Brain Res 134: 541–545

Gramsbergen A (1976) EEG development in normal undernourished rats. Brain Res 105: 287–308

Gray EG (1971) The fine structural characterization of different types of synapse. Prog Brain Res 34: 149–160

Grunewald W (1969) Digitale Simulation eines räumlichen Diffusionsmodells der O_2-Versorgung biologischer Gewebe. Pflügers Arch 309: 266–284

Guseo A, Gallyas F (1974) Intercapillary bridges and the development of brain capillaries. In: Cervos-Navarro J (ed) Pathology of cerebral microcirculation. de Gruyter, Berlin New York, pp 448–453

Gyllensten L (1959) Postnatal development of the visual cortex in darkness (mice). Morphol Neerl Scand 2: 331–345

Hasegawa T, Ravens JR, Toole JF (1967) Precapillary arteriovenous anastosomes. "Thoroughfare channels" in the brain. Arch Neurol 16: 217–224

Hauck G (1971) Organisation und Funktion der terminalen Strombahn. In: Bauereisen E (ed) Physiologie des Kreislaufs 1. Springer, Berlin Heidelberg New York, pp 99–144

Hauck G (1975) Capillary bed architecture and perfusion pattern. Drug Res 25: 5

Haug H (1975) Neuere Aspekte über den biologischen Alterungsvorgang im menschlichen Gehirn. Verh Anat Ges 69: 389–395

Haug H, Kölln M, Rast A (1976) The postnatal development of myelinated nerve fibres in the visual cortex of the cat. A stereological and electron microscopical investigation. Cell Tissue Res 167: 265–288

Hazama F, Haebara H, Amano S, Ozaki T (1977) Autoradiographic investigation of cell proliferation in the brain of spontaneously hypertensive rats. Acta Neuropathol (Berl) 37: 231–236

Hedner Th (1978) Central monoamine metabolism and neonatal oxygen deprivation. An experimental study in the rat brain. Acta Physiol Scand (Suppl) 460: 1–34

Henn FA, Haljamäe H, Hamberger A (1972) Glial cell function: Active control of extracellular K+ concentration. Brain Res 43: 437–443

Heubner O (1874) Die luetische Erkrankung der Hirnarterien. Leipzig

Hinds JW, McNelly NA (1977) Aging of the rat olfactory bulb: growth and atrophy of constituent layers and changes in size and number of mitral cells. J Comp Neurol 171: 345–368

Hirsch H, Schneider M (1968) Kapillarisierung des Gehirns. In: Hirsch H, Schneider M (eds) Handbuch der Neurochirurgie. Grundlagen II. Springer, Berlin Heidelberg, New York

Horstmann E (1959) Die postnatale Entwicklung der Kapillarisierung im Gehirn eines Nesthockers (Ratte) und eines Nestflüchters (Meerschweinchen). Anat Anz 106/107: 405–410

Horstmann E (1960) Abstand und Durchmesser der Kapillaren im Zentralnervensystem verschiedener Wirbeltierklassen. In: Tower DB, Schade JP (eds) Structure and function of the cerebral cortex. Elsevier, Amsterdam, pp 59–63

Illig L (1961) Pathologie und Klinik in Einzeldarstellungen: Bd X, Die terminale Strombahn Springer, Berlin Göttingen Heidelberg

Ismailowa IW (1958) Die Angioarchitektonik der Großhirnrinde des Menschen. Sowjetwissenschaft, Naturwissenschaftl Beiträge 8: 839–847 Verlag Kultur und Fortschritt, Berlin

Iwanowski L (1974) The role of connective tissue in the brain aging. VII Int Congr Neuropath Budapest Akad Kiadó, Budapest, p 293

Johannesson G, Brun A, Gustafson L, Engvar D (1977) EEG in presenile dementia related to cerebral blood flow and autopsy findings. Acta Neurol Scand 56: 83–103

Källén B, Valmin K (1960) Morphogenetic aspects on the alkaline phosphatase distribution in embryonic chick brains. Z Anat Entwicklungsgesch 121: 376–387

Kaplan HA, Ford DH (1966) The brain vascular system. Elsevier, Amsterdam

Kapustina EV (1952) Entwicklung des Arteriennetzes in der Pia mater der Großhirnhemisphäre menschlicher Föten in der zweiten Hälfte des intrauterinen Lebens (Übersetzung). Pediatriya (Mosc) 5: 30–38

Kennedy C, Grave GD, Jehle JW, Sokoloff L (1970) Blood flow to white matter during maturation of the brain. Neurology (Minneap) 20: 613–618

Kennedy C, Grave GD, Jehle JW, Sokoloff L (1972) Changes in blood flow in the component structures of the dog brain during postnatal maturation. J Neurochem 19: 2423–2433

Knobloch J (1956) Alterswandlungen in der Reaktionsfähigkeit des feinsten Gefäßabschnittes beim Menschen. Ph Thesis Leipzig

Kontos HA, Wei EP, Raper AJ, Rosenblum WJ, Navari RM, Patterson JL (1978) Role of tissue hypoxia in local regulation of cerebral microcirculation. Am J Physiol 234: H 582–H 591

Korr H, Schlutze B, Maurer W (1975) Autoradiographie investigations of glial proliferation in the brain of adult mice. II. Cycle time and mode of proliferation of neuroglia and endothelial cells. J Comp Neurol 160: 477–490

Krebs W, David H (1962) Beitrag zur elektronenmikroskopischen und funktionellen Altersveränderungen der Kapillaren. Dtsch Gesundheitswes 17: 1845–1849

Land JM, Booth RFG, Berger R, Clark JB (1977) Development of mitochondrial energy metabolism in rat brain. Biochem J 164: 339–348

Lazorthes G, Espagno J, Lazorthes Y, Zadeh JO (1968) The vascular architecture of the cortex and the cortical blood flow. In: Luyendijk W (ed) Cerebral circulation. Elsevier, Amsterdam (Prog Brain Res, Vol 30, pp 27–32)

Lewis FT (1933) Mathematically precise features of epithelial mosaics: observations on the endothelium of capillaries. Anat Res 55: 323–341

Lierse W (1963) Über die Beeinflussung der Hirnangioarchitektur durch die Morphogenese. Acta Anat 53: 1–54

Lübbers DW (1968) The oxygen pressure field of the brain and its significance for the normal and critical oxygen supply of the brain. In: Lübbers DW, Luft UC, Thews G, Witzleb E (eds) Oxygen transport in blood and tissue. Thieme, Stuttgart, pp 124–139

Malpighi M (1661) De pulmonibus observationes anatomicae. Epistula secunda ad Borellum, Bologna

Mareš P, Vítová Z (1973) Development of cortical visual evoked potentials in the rat. Physiol bohemoslov (Engl ed Cesk Fysiol) 22: 469–476

Mareš VL, Lodin Z (1974) An autoradiographic study of DNA synthesis in adolescent and adult mouse forebrain. Brain Res 76: 557–561

Masters JRW (1976) Epithelial-mesenchymal interaction during lung development: The effect of mesenchymal mass. Develop Biol 51: 98–108

Mayrovitz HN, Wiedeman MP (1975) Microvascular hemodynamic variations accompanying microvessel dimensional changes. Microvasc Res 10: 322–339

Menkes B, Alexandru C, Tudose O, Checiu I (1975) Untersuchungen über die Stratigenese im Zentralnervensystem. Z Mikrosk Anat Forsch (Leipz) 89: 63–78

Müller E, Pearse AGE, Moss DW (1971) The inducible alkaline phosphatase of rat heart. Some properties of the enzyme and factors influencing its activity. Biochem J 123: 895–900

Nag S, Robertson DM, Dinsdale HB (1977) Cerebral cortical changes in acute experimental hypertension: an ultrastructural study. Lab Invest 36: 150–161

Nanda BS, Getty R (1971) Age related histomorphological changes in the cerebral arteries of domestic pig. Exp Gerontol 6: 453–460

Niessing K (1950) Zur funktionellen Histologie der Hirnkapillaren. Verh Anat Ges 48: 42–60

Ono T, Cutler RG (1978) Age-dependent relaxation of gene repression-Increase of endogenous murine leukemia virus-related and globin-related RNA in brain and liver of mice. Proc Natl Acad Sci USA 75: 4431–4435

Opitz E, Schneider M (1950) Über die Sauerstoffversorgung des Gehirns und den Mechanismus von Mangelwirkungen. Ergeb Physiol 46: 125–260

Panula P, Rechardt L (1978) Age-dependent increase in the non-specific cholinesterase activity of the capillaries in the rat neostriatum. Histochemistry 55: 49–54

Pentschew A, Garro F (1966) Lead encephalo-myelopathy of the suckling rat and its implications on the porphyrinopathic nervous diseases. With special reference to the permeability disorders of the nervous system's capillaries. Acta Neuropathol (Berl) 6: 266–278

Petito CK, Schaefer JA, Plum F (1977) Ultrastructural characteristics of the brain and blood-brain barrier in experimental seizures. Brain Res 127: 251–267

Petrén T (1938) Untersuchungen über die relative Kapillarlänge der motorischen Hirnrinde in normalem Zustande und nach Muskeltraining. Anat Anz 85: (Ergebnis Heft) 169–172

Pfeifer RA (1928) Die Angioarchitektonik der Großhirnrinde. Springer, Berlin

Pfeifer RA (1931) Anastomosen der Hirngefäße. J Psychol Neurol 42: 1–173

Pfeifer RA (1940) Die angioarchitektonische areale Gliederung der Großhirnrinde. Thieme, Leipzig

Pysh JJ (1970) Mitochondrial changes in rat inferior colliculus during postnatal development: An electron microscopic study. Brain Res 18: 325–342

Rhodin JAG (1967) The ultrastructure of mammalian arterioles and precapillary sphincters. J Ultrastruct Res 18: 181–223

Rhodin JAG (1968) Ultrastructure of mammalian venous capillaries, venules, and small collecting veins. J Ultrastruct Res 25: 452–500

Richardson KL, Jarett L, Finke EH (1960) Embedding in epoxy resins for ultrathin sectioning in electron microscopy. Stain Technol 35: 313–323

Rickmann M, Wolff JR (1976) On the earliest stages of glial differentiation in the neocortex of rat. Exp Brain Res (Suppl) I: 239–243

Risberg J, Ancri A, Ingvar DH (1969) Correlation between cerebral blood volume and cerebral blood flow in the cat. Exp Brain Res 8: 321–326

Rostan L (1823) Recherches sur le ramollissement du cerveau. Paris

Samaras TT (1974) The law of entropy and the aging process. Hum Dev 17: 314–320

Saunders RL, Bell MA (1971) X-ray microscopy and histochemistry of the human cerebral blood vessels. Special Report: Cerebral microcirculation. J Neurosung 35: 128–140

Scharrer E (1944) The capillary bed of the central nervous system of certain invertebrates. Biol Bull 87: 52–58

Scharrer E (1945) Capillaries and mitochondria in neuropil. J Comp Neurol 83: 237–243

Scharrer E (1962) Brain function and the evolution of cerebral vascularization. In: James Arthur lecture on the evolution of the human brain, pp. 1–32, The American Museum of Natural History, New York

Scheibel ME, Lindsay RD, Tomiyasu U, Scheibel AB (1975) Progressive dendritic changes in aging human cortex. Exp Neurol 47: 392–403

Schöller P (1972) Zentralitätsforschung. Wissenschaftliche Buchgesellschaft Darmstadt

Sessa G, Perez MM (1975) Biochemical changes in rat brain associated with the development of the blood-brain barrier. J Neurochem 25: 779–782

Shefer VF (1973) Absolute number of neurons and thickness of the cerebral cortex during aging, senile and vascular dementia, and Pick's and Alzheimer's diseases. Neurosci Behav Physiol 6: 319–324

Sheldrake AR (1974) The ageing, growth and death of cells. Nature 250: 381–385

Silini G, Andreozzi U (1974) Haematological changes in the ageing mouse. Exp Gerontol 9: 99–107

Smith CG (1934) The volume of the neocortex of the albino rat and the changes it undergoes with age after birth. J Comp Neurol 60: 319–347

Sokoloff L (1977) Relation between physiological function and energy metabolism in the central nervous system. J Neurochem 29: 13–26

Spalteholz W (1923) Gefäßbaum und Organbildung. Arch Entwicklungsmech Org (Wilhelm Roux) 52–97: 480–531

Strong LH (1961) The first appearance of vessels within the spinal cord of the mammal: Their developing patterns as far as partial formation of the dorsal septum. Acta Anat 44: 80–108

Strong, LH (1964) The early embryonic pattern of internal vascularization of the mammalian cerebral cortex. J Comp Neurol 123: 121–138

Tomasch J (1972) Gibt es einen altersbedingten kontinuierlichen Neuronenverlust? Wien Klin Wochenschr 84 169–173

Tyler DB, Harreveld van A (1942) The respiration of the developing brain. Am J Physiol 136: 600–603

Vaughan DW, Peters A (1974) Neuroglial cells in the cerebral cortex of rats from young adulthood to old age: an electron microscope study. J Neurocytol 3: 405–429

Vaughan WJ, Calvin M (1977) Electrophoretic analysis of brain proteins from young adult and aged mice. Gerontology 23: 110–126

Verley R, Axelrad H (1975) Postnatal ontogenesis of potentials elicted in the cerebral cortex by afferent stimulation. Neurosci Letters 1: 99–104

Vernadakis A (1975) Neuronal-glial interactions during development and aging. Fed Proc 34: 89–95

Vicq d'Azyr M (1786) Traité d'anatomie et de physiologie. Paris

Vital-Durand F, Garey LJ, Blakemore C (1978) Monocular and binocular deprivation in the monkey: morphological effects and reversibility. Brain Res 158: 45–64

Vrensen G, de Groot D, Nunes-Cardozo J (1977) The postnatal development of neurons and synapses in the visual and motor cortex of rabbits: a quantitative light and electron microscopic study. Brain Res Bull 2: 405–416

Vuia O (1966) Intracerebral angioarchitectonic venous territories (their role in venous draining of brain under normal and pathologic conditions). Res Roum Neurol 3: 235–243

Warshaw JB, Terry ML (1976) Cellular energy metabolism during fetal development. VI. Fatty acid oxidation by developing brain. Dev Biol 52: 161–166

Weibel ER (1969) Stereological principles for morphometry in electron microscopic cytology. Int Rev Cytol 26: 235–302

Weiss HR, Edelman NH (1976) Effect of hypoxia on small vessel blood content of rabbit brain. Microvasc Res 12. 305–316

Westergaard E (1974) Transport of protein tracers across cerebral arterioles under normal conditions. In: Cervos-Navarro J (ed) Pathology of cerebral microcirculation. de Gruyter, Berlin New York, pp 218–227

White P, Hiley CR, Goodhardt MJ Carrasco LH, Kett JP Williams IEI, Bowen DM (1977) Neocortical cholinergic neurons in elderly people. Lancet 7974: 668–671

Wiederhold K-H, Bielser JrW, Schulz U, Veteau M-J, Hunziker O (1976) Three dimensional reconstruction of brain capillaries from frozen serial sections. Microvasc Res 11: 175–180

Wolff JR (1976a) An ontogenetically defined angioarchitecture of the neocortex. Drug Res 26: 1246

Wolff JR (1976b) Quantitative analysis of topography and development of synapses in the visual cortex. Exp Brain Res (Suppl.) 1: 259–263

Wolff JR (1978) Ontogenetic aspects of cortical architecture: Lamination. In. Brazier MAB, Petsche H (eds) Architectonics of the cerebral cortex. Raven Press, New York, pp 159–173

Wolff JR, Bär Th (1976) Development and adult variations of the pericapillary glial sheath in the cortex of rat. In: Cervos-Navarro J, Matakas F (eds) The cerebral vessel wall. Raven Press, New York, pp 7–13

Wolff JR, Goerz Ch, Bär Th, Güldner FH (1975) Common morphogenetic aspects of various organotypic microvascular patterns. Microvasc Res 10: 373–395

Woossmann H, Kreher AC, Nitschkoff S (1977) Changes of activity of alcaline phosphatase of the circulatory system in experimental hypertension of the rat. Acta Histochem 58: 11–16

Subject Index

Other Reviews of Interest in this Series

Part 6: **Lüdicke, M.**: Internal Ear
Angioarchitectonic of Serpents.
21 figures. 41 pages. 1978.
ISBN 3-540-08836-9

Volume 55

Part 1: **Reutter, K.**: Taste Organ
in the Bullhead (Teleostei).
20 figures. 98 pages. 1978.
ISBN 3-540-08880-6

Part 2: **Dvorák, M.**: The
Differentation of Rat Ova
During Cleavage. 62 figures.
131 pages. 1978.
ISBN 3-540-08983-7

Part 3: **Wagner, H.-J.**: Cell Types and
Connectivity Patterns in Mosaic Retinas.
30 figures. 81 pages. 1978.
ISBN 3-540-09013-4

Part 4: **Jones, D.G.**: Some Current
Concepts of Synaptic Organization.
21 figures. 69 pages. 1978.
ISBN 3-540-09011-8

Part 5: **Fleischer, G.**: Evolutionary
Principles of the Mammalian Middle
Ear. 25 figures. 70 pages. 1978.
ISBN 3-540-09140-8

Volume 56

Kaissling, B.; Kriz, W.:
Structural Analysis of the Rabbit Kidney.
47 figures. VIII, 123 pages. 1979.
ISBN 3-540-09145-9

Volume 57

Niimi, K., Matsuoka, H.:
Thalamocortical Organization of the
Auditory System in the Cat Studied by
Retrograde Axonal Transport of Horse-
radish Peroxidase. 30 figures. X, 56
pages. 1979.
ISBN 3-540-09449-0

Volume 58

Verwoerd, C.D.A. van Oostrom, C.G.:
Cephalic Neural Crest and Placodes.
41 figures. VI, 75 pages. 1979.
ISBN 3-540-09608-6

Springer-Verlag Berlin Heidelberg New York